45 Advances in Biochemical Engineering
Biotechnology

Managing Editor: A. Fiechter

W0107640

Enzymes and Products from Bacteria Fungi and Plant Cells

With contributions by
T. Coolbear, R. M. Daniel, C. P. Kubicek,
P. K. R. Kumar, H. W. Morgan, K. Schügerl,
A. Singh, D. C. Taylor, E. W. Underhill,
N. Weber

With 19 Figures and 29 Tables

Springer-Verlag
Berlin Heidelberg GmbH

ISBN 978-3-662-14993-5   ISBN 978-3-540-46725-0 (eBook)
DOI 10.1007/978-3-540-46725-0

© Springer-Verlag Berlin Heidelberg 1992

Originally published by Springer-Verlag Berlin Heidelberg New York in 1992
Softcover reprint of the hardcover 1st edition 1992

Library of Congress Catalog Coard Number 72-152360
Printed of Germany

Typesetting: Th. Müntzer, Bad Langensalza;

02/3020-5 4 3 2 1 0 – Printed on acid-free paper

Table of Contents

The Cellulase Proteins of *Trichoderma reesei*: Structure, Multiplicity, Mode of Action and Regulation of Formation

Christian P. Kubicek

Abteilung für Mikrobielle Biochemie, Institut für Biochemische Technologie und Mikrobiologie, Technische Universität Wien, Getreidemarkt 9, 1060 Wien, Austria

The filamentous fungus *Trichoderma reesei* is the predominant industrial producer of cellulolytic enzymes by secreting an enzyme system capable of degrading crystalline cellulose, which consists of several cellobiohydrolases, endoglucanases and β-glucosidases. All of these enzymes occur in multiple forms. A critical appraisal of the methods used to assess cellulase multiplicity is presented. By the aid of gene technology, advanced protein analytics and immunology, "true" isoenzymes and proteolytic fragments of all of these enzymes could be identified, and their structure and properties are described. Also, the recent elucidation of the three-dimensional domain structure of cellulases, their active center, and the role of both in the hydrolysis of cellulose are dealt with. Particular emphasis is presented on the differences in the enzymatic reaction mechanisms of cellobiohydrolase I and II, and their synergism.

Recent developments in the understanding of the triggering of cellulase formation by cellulose and its inhibition by readily metabolizable carbon sources are also presented.

Advances in Biochemical Engineering/
Biotechnology, Vol. 45
Managing Editor: A. Fiechter
© Springer-Verlag Berlin Heidelberg 1992

1 Introduction

Cellulose is by far the most abundant renewable carbohydrate source with an estimated synthesis rate of 4×10^7 tons per year. Many attempts have been made to utilize this enormous amount through enzymatic hydrolysis into glucose. Cellulolytic enzyme mixtures may be obtained from several microorganisms, but due to the physical nature of the cellulose molecule and the fact that its natural occurrence is always accompanied by associated materials (hemicelluloses, lignin), their efficiency deserves improvement in most cases, and different enzymes may be preferred for different purposes. However, the inherent complexity of cellulolytic enzyme mixtures in terms of number of enzyme components and their specificity has caused a still prevailing lack of sufficient understanding. This situation is still valid even for the most extensively studied cellulolytic microorganism, the fungus *Trichoderma reesei* and its mutant strains. As I will explain later, this is in part due to the difficulties to purify some of these enzymes to homogeneity. Considerable progress in our understanding of *T. reesei* cellulases has however recently been provided by studies involving genetic engineering techniques, monoclonal antibodies and physicochemical protein analytics (for Refs. see below). It is therefore the purpose of this paper to critically assess the current state of knowledge about the nature of the *T. reesei* enzymes involved in cellulose breakdown, and to define important areas still unsufficiently understood. It is noted that related aspects as conditions for production and application of the enzymes, albeit equally important, will not be covered here. These topics have already been subject of several detailed reviews recently [1–5].

2 General Molecular Properties
of *T. reesei* Cellulolytic Enzymes

Cellulose is a linear β-1,4-glucosidically linked homopolymer of around 8000–12000 glucose units, which forms a crystalline unit held together by hydrogen bonding [6]). According to this structure, Mandels and Reese [7] postulated the involvement of two different types of enzymes in the degradation of natural cellulose: a "C_1"-enzyme, which renders the cellulose crystal accessible for hydrolytic attack, and a "C_x"-enzyme, which subsequently degrades cellulose by both endo- as well as exo-type attack. As will be reviewed in chapter 6, evidence for the existence of the C_1-enzyme is still obscure, but at least two exocellobiohydrolases (EC 3.2.1.91, CBH), several endo-β-1,4-glucanases (EC 3.2.1.4, EG), and β-glucosidases (EC 3.2.1.21) have been identified and characterized.

Enzymes acting on molecules like cellulose, which are insoluble and of similar, or even greater size than themselves, obviously require delicate tertiary structures: comparison of the sequence of approximately 50 fungal and bacterial cellulase genes and of some other polysaccharide hydrolase (for review see [8]) has offered evidence that such proteins are composed of separate domains, which allow a spacial separation of the sites involved in substrate recognition and enzymatic activity.

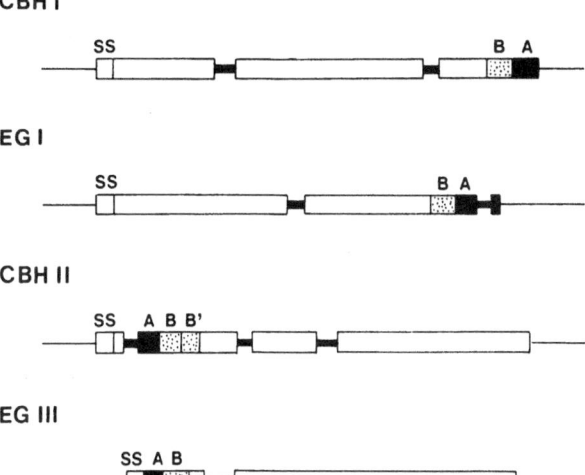

Fig. 1. Domain structure of the four major cellulase genes from *T. reesei*. Taken from Ref. [101], by permission

First evidence for a domain organization of the *T. reesei* cellulases came from the finding that partial proteolysis by papain can cleave CBH I and CBH II into an enzymatic active "core", and a cellulose adsorbing "tail" domain [9, 10]. These studies were complemented by data on the gene sequences of four main cellulase components from *T. reesei* (CBH I, CBH II, EG I, EG III), which revealed a strikingly conserved terminal domain in all four species [11], which is joined to the rest of the protein by a similarly conserved "hinge" domain (Fig. 1).

Coinciding evidence for such a domain organization of CBH I and II came from the pioneering work of I. Pilz and coworkers using small-angle X-ray scattering [12, 14]. These studies revealed a rather unusual tadpole like shape of both CBH I and II, with isotropic heads and protruding tails (Fig. 2). Corresponding data on the size estimates are given in Table 2. Due to the duplication of the B region in CBH II, this molecule is somewhat longer despite its lower molecular mass. Crystallization of either of these enzymes or any other cellulase has not yet been possible. However, after proteolytic removal of the AB-region, CBH II was recently crystallized [15]. This indicates that the A-, B-, or AB-region provides some flexibility to the CBH molecules. In Pettersons' group, AB-domains of CBH I from *T. reesei* and *P. chrysosporium* were isolated by proteolytic cleavage, and compared to a synthetic peptide synthesized according to the residues 462–497 of *T. reesei* CBH I. They were able to show [16] that the synthetic sequence (which lacks the carbohydrate rich B region) binds equally well to cellulose, but requires intact disulfide bonds. These studies support the idea that the AB-domain forms a functional domain wherein the A part interacts with cellulose, whereas the B-region provides a flexible arm connecting the catalytic and the adsorptive regions of the enzyme. The structure of this synthetic peptide in solution has more recently

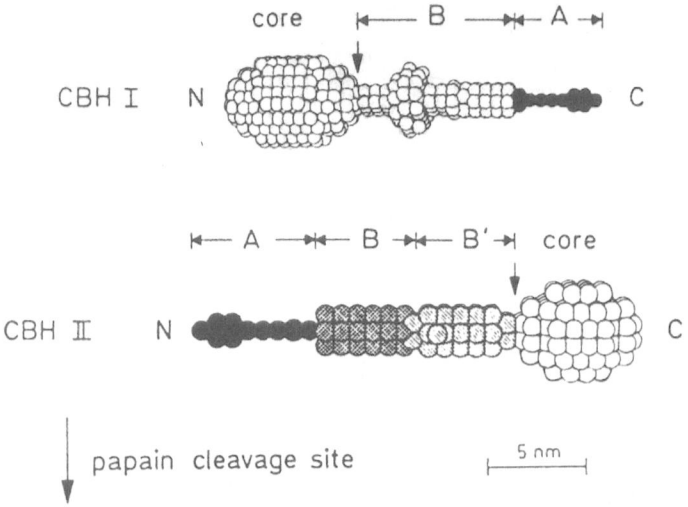

Fig. 2. Model structure of CBH I and CBH II from *T. reesei* taken from Ref. [14], by permission

been investigated by NMR and hybrid distance geometry-dynamical simulated annealing [17]. It was found to be made up of an irregular triple-stranded antiparallel β-sheet (β1, β2, β3), in which β3 is hydrogen bonded to the two other sheets. Its three-dimensional structure exhibits a wedgelike shape with an amphi-philic character, one face being predominantly hydrophilic, and the other hydro-phobic. Both faces would be ideally suited to interact with cellulose, which has a flat, layered structure with hydrogen bondings in the plane of the layer and Van der Waals interactions holding the layer together [18].

The B-region carries heavy O-glycosylation, but the reason for this is still speculative. It has been assumed that it may serve to protect the enzymes against proteolysis [16]; however, proteolytic cleavage in fact starts by attacking within the B-region [10].

Detailed knowledge on the three-dimensional structure of cellulases has been strongly hampered by the failure to crystallize these proteins, however, recently crystallization of the catalytic core proteins of CBH II (and CBH I) has been successful [15]. CBH II consists of a seven strand singly wound parallel α/β-barrel (Fig. 3). Extended loops from the barrel produce a large channel for cellulose binding. The active site is located at the C-terminus of the β-sheet and can be clearly identified using data collected with an inhibitor diffused into the crystal. It is present in the tunnel, through which the cellulose threads, and two aspartic acid residues most probably form the catalytic residues (see also chapter 6). In this structure, the ABB' block (which is about 140 A long [14], see Table 1) is only 45 A apart from the entrance of the tunnel. It is therefore tempting to speculate that the anchoring non-catalytic domain holds the cellulose crystal in the appropriate position for the active center to attack.

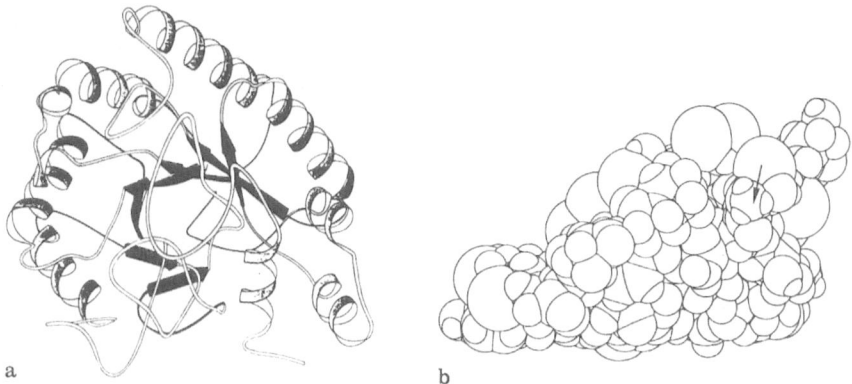

Fig. 3a, b. Schematic representation of the crystal structure of *T. reesei* CBH II core protein. Taken from Ref. [199], by permission

Although corresponding data for CBH I is still lacking, the structure outlined above may be typical for cellobiohydrolases, i.e. exoglucanases. Comparable investigations for endoglucanases are also still lacking, but a comparison of respective amino acid sequences (derived from gene sequences) show that e.g. EG I shows four clear deletions relative to CBH I; these deletions result in a failure to form the tunnel and probably result in a groove for multiple subsites [17]. Kraulis et al. [17] supposed that the various cellulases secreted by *T. reesei* may differ in the number of surface loops forming the active site. Moreover, the tunnel-like structure may be typical for exoglucanases, since it allows an attack only from the end of the cellulose chain.

Being secreted proteins, most of the cellulases from *T. reesei* have proven to be glycoproteins. Their carbohydrate structure may be deduced from detailed analysis with CBH I and in part EG I: it is composed both of N- as well as O-linked chains; most of the O-glycosylation occurs in the B ("hinge") region of the protein, whereas N-glycosylation is restricted to the core domain, most

Table 1. Concentrations of individual cellulases in the culture fluid upon cultivation of *T. reesei* QM 9414 under different conditions*

Conditions	Cellulase concentration (mg l⁻¹)			
	CBH I	CBH II	EG I	β-Gluc
Glucose	<5	12	<5	7
Cellulose	960	270	250	12
Lactose	231	69	87	8
Sophorose	140	25	15	<5

* Data are taken from Ref. [66], by permission; β-glucosidase data (β-gluc) are from unpublished results in the authors laboratory

probably at β-turns. Such a location would be consistent with its protective role against proteolysis [19]. The N-linked carbohydrate resembles the highly conserved (Man)$_9$(ManNAc)$_2$ structure, found in other organisms, but some trimmed (Man)$_5$(GlcNAc)$_2$ antennae were also found [20]. In CBH I, three-N-glycosylation sites are occupied, but the proportion of Man$_9$ and Man$_5$ structures is roughly 1 : 4, hence suggesting microheterogeneity [20]. Carbohydrate microheterogeneity of CBH I has also been reported by Gum and Brown [21], yet these authors were unable to detect the N-glycosylation of CBH I. On the other hand, PSMS analysis of CNBr-peptides from CBH I confirmed the presence of 3-N-glycosylation sites, but failed to reveal any heterogeneity [21, 22].

The situation with O-glycosylation is even less satisfying: since this type of glycosylation lacks a defined target sequence, no predictions from the aa-sequence are possible. Analyses by different authors [20, 23, 24] revealed the presence of mono-, di-, tri- and tetrasaccharide chains. Attachment of these oligosaccharides to protein has also been observed in microsomes of *T. reesei* [25]. The relative proportions of these oligosaccharides in CBH I, however, differed strongly in different reports. Heterogeneity of O-Glycosylation of the CBH I-AB-region was analyzed by NMR [26], thereby showing that one enzyme population carries 24, and another one only 19 mannose units. The function of O-Glycosylation was discussed with respect to protection against proteolyses or adsorption to cellulose, but current knowledge for this is rather contradictory. On the other hand, there is some evidence, that O-glycosylation may be required for efficient secretion of cellulases [27, 28].

3 Enzyme Composition of *T. reesei* Cellulase Preparations — A Survey of Methods and Tools

While considerable progress has so far been obtained in the understanding of the structure and function of a few individual cellulases, the number and types of cellulases present in *T. reesei* are still a matter of dispute. This is in part caused by the fact that the method of their assessment has so far often been the stumblestone in the identification of the number of enzymes, which are involved in cellulose breakdown by *T. reesei*.

Electrophoretic techniques, especially IEF, have been frequently applied to examine the cellulase spectrum secreted by *T. reesei* [29, 32]. Identification of cellulases within the separated proteins can be conveniently achieved by overlaying the gels with cellulose followed by Congo Red [33] or dyed cellulose [34], and differentiation of different cellulases may be acchieved by staining with specific chromogenic artificial substrates [35]. Results obtained by IEF lead to the believe that the number of cellulase isoenzymes is very high [36, 37]. IEF is usually a powerful tool to separate isoenzymes with very similar physicochemical properties, and some of the multiple cellulases detected thereby may in fact be the result of carbohydrate heterogeneity [20, 21, 26], partial proteolysis [38, 39], transcript heterogeneity [40] or other processes. IEF, however, is also subject to some pitfalls: firstly, ampholines may complex with some proteins and thereby produce multiple bands [41]; some cellulases may also form very stable complexes with each other: Sprey has shown that single bands of cellulases in IEF reflect the purity of (multi)enzyme complexes rather than that of the pure enzymes [42–45]. The forces involved in the association of cellulase proteins are unclear: some cellulases contain strongly hydrophobic domains, which may lead to their interaction with each other in aqueous environment [46, 47]. Sprey favored the role of an acidic, cell-wall derived polysaccharide, which may bind different cellulases [43], and hence lead to complex formation. Evidence for this polysaccharide has so far not been presented, but Sprey cites unpublished data that it contains „acidic carbohydrate chains". We have recently purified and characterized an acidic heteroglycan from *T. reesei* [48]. However, its role as an anchor for cellulase complex formation has yet to be assessed. In practice, all these factors may contribute to one or other extent to the multitude of bands seen in IEF, and therefore, render this technique of doubtful value in the analysis of cellulases.

Chromatographic techniques, particularly in combination with HPLC or FPLC, have also frequently been used to analyze the composition of cellulase enzymes [49–55]. Basically, these methods suffer from the same disadvantages as described above for IEF techniques. However, higher amounts of protein can be applied, and critical peaks can therefore be re-investigated by other techniques, or checked for cross-contamination by other enzymes [55].

Immunological techniques, using polyclonal antibodies, were introduced into cellulase analytics very early [56–58], but their specificity and hence the result obtained was always a matter of dispute [59, 60]. The specificity of polyclonal antibodies, raised against CBH I, CBH II and EG I, among which each polyclonal antiserum showed cross-reactivity with other cellulases, has recently been analyzed in detail [61]: cDNAs lacking regions coding for certain functional domains were produced by preparing series of 3'-end deletions, and the corresponding truncated proteins were obtained by expressing the cDNAs in yeast. The corresponding Western blots showed that all antibodies were almost entirely directed against the conserved terminal regions of the cellulase enzymes (see later for their description). This clearly emphasises caution when quantitative immunological assays are to be used to analyze the contents of individual cellulases.

Monoclonal antibodies clearly do not show this disadvantage, since − despite of high homology between some cellulases − some epitopes exist which are not shared by different cellulase enzymes. The isolation and properties of monoclonal antibodies against CBH I, CBH II, EG I, EG III and β-glucosidase have now been described [46, 47, 62, 63] and their application in cellulase quantitation by dot-blot-scanning [64, 65] and ELISA [66, 67] has been reported. The availability of monoclonal antibodies against different epitopes of CBH I and II [62] also offers the quantitation of intact and truncated forms of these enzymes [66]. Interestingly, some cellulase preparations contain unknown components interfering with ELISA, which have to be removed by precipitation of the cellulases with ethanol [66] or heat-treatment [67]. The cellulase composition of different *T. reesei* culture fluids, analyzed by ELISA, is given in Table 2.

Table 2. Molecular dimensions of CBH I, CBH II and their cores

Domain		Intact [nm]	Core [nm]	Tail [nm]
Head	CBH I			
	l	6.7	6.7	12.9
	d	4.4	4.5	3.2
	CBH II			
	l	5.4	6.0	15.2
	d	5.0	5.0	3.5

Data taken from Ref. [14], by permission

4 Properties of Individual Cellulases

4.1 Cellobiohydrolases

Cellobiohydrolases, as explained earlier, are defined as enzymes which split off cellobiose units from the non-reducing end of the chain. Two such enzymes, each occurring in several isoenzymic forms, have so far been identified in *T. reesei* − CBH I and CBH II − which are dealt with below:

4.1.1 Cellobiohydrolase I

CBH I comprises the major part of the cellulolytic enzyme mixture secreted by *T. reesei* (cf. Table 2). Its gene has been cloned and sequenced [69, 70]. It codes for a polypeptide with a corresponding M_r of 58 kDa. As emphasized earlier, it is both N- and O-glycosylated. Data from gene sequencing was completely consistent with the primary structure of CBH I protein by automatic liquid phase Edman degradation [71], when two different mutants were compared (L27 and QM 9414). However, a comparison of the gene sequence of *T. reesei* L27 and *T. viride* indicates only 95% homology, hence indicating several differences in the aa sequence of CBH I purified from different *Trichoderma* species.

Purification of CBH I has been reported by several authors [72–80]. The corresponding data is compiled in Table 3, which indicates considerable differences in the M_r of the purified proteins. However, the relative proportions of amino acids present in these proteins are well comparable. It therefore appears that the molecular weight determination of cellulases is subject to severe pitfalls. In any case, from this data there is little evidence for more than one single CBH I enzyme secreted by *T. reesei*.

Purified CBH I exhibits pronounced heterogeneity, but the reason for this has so far not been assessed clearly: early evidence of carbohydrate heterogeneity [21] has more recently been specified as O- and N-linked carbohydrate microheterogeneity [20, 26]. However, differential transcription termination has also been

Table 3. Amino acid composition and some properties of CBH I isolated from *T. reesei*

Ref.	[79]	[59]	[21]	[78]	[73]	[77]
Cys	4.8	3.8	4.1	4.7	4.3	4.1
Asx	11.3	11.9	12.5	12.4	12.0	11.2
Thr	11.5	11.3	10.7	12.7	10.7	11.3
Ser	11.3	10.8	10.4	10.8	11.2	11.3
Glx	8.2	9.0	8.2	10.0	8.8	8.2
Pro	5.6	5.6	6.4	5.0	5.7	6.0
Gly	12.5	13.3	13.2	15.1	12.3	12.7
Ala	5.8	6.3	6.8	7.1	6.0	6.2
Val	4.6	5.1	4.9	5.3	4.8	4.5
Met	1.2	1.4	1.6	1.5	1.1	1.4
Ile	2.4	2.4	2.3	2.1	2.2	1.9
Leu	5.6	5.9	5.5	6.1	5.8	5.2
Tyr	4.8	5.0	5.0	5.0	5.2	5.1
Phe	3.0	3.2	3.0	3.4	2.5	3.1
His	1.0	1.1	1.0	1.0	1.0	0.9
Lys	2.6	2.8	2.5	2.9	2.7	2.6
Trp	1.8	ND	ND	2.4	1.6	1.5
Arg	1.8	1.9	1.7	2.1	1.6	2.2
[kDa]	64.0	66.0	48.3	50.3	42.0	61.0*
CH [%]	6.0	ND	10.4	7.2	9.2	7.0

All amino acid values are given as % of total residues; ND, not determined; [kDa] indicates the M_r; CH [%] the carbohydrate content; * value for protein lacking carbohydrate given

observed [40]. Furthermore, tight binding of oligosaccharides from the medium to CBH I during purification has been reported [81]. All these observations may be the result of a very relaxed mechanism for glycoprotein formation and processing in *T. reesei*. However, multiplicity of CBH I does not appear randomly during cultivation, suggesting that it is an inherent property of this enzyme or its formation. It remains to be assessed whether enzyme microheterogeneity in fact fulfills a yet unknown physiological function.

4.1.2 Cellobiohydrolase II

The purification and characterization of the second cellobiohydrolase secreted by *T. reesei* was first described as an enzyme immunologically distinct from CBH I, which produced cellobiose from cellulose [57]. However, for a long time its purification presented a major methodological problem because its physico-chemical properties (IEP, M_r) are very similar to EG I. This was finally overcome by using immunoadsorption [82] or affinity chromatography on thiocellobiose coupled to Affigel [83]. The *cbh 2* gene has been cloned and sequenced [84, 85]. It codes for a 471 aa protein, whose aa composition is identical to that of purified CBH II [79]. It contains a duplicated B-motive, and therefore, is glycosylated stronger than CBH I. Probably owing to the paucity of reports on purification of CBH II, no multiple forms have yet been described.

4.2 Endoglucanases

Endoglucanases hydrolyze β-1,4-glycosidic linkages randomly. They do not attack cellobiose but hydrolyze cellodextrins, phosphoric-acid swollen cellulose and substituted celluloses such as carboxymethyl (CM)- and hydroxyethyl (HE)-cellulose. *Trichoderma* seems to secrete a number of respective enzymes into the medium, of which two (EG I, EG III) have been the subject of a more detailed investigation. It should be noted that some confusion has arisen in the past from the fact that EG III was termed EG II by several authors from the US. However, this has been overcome now.

4.2.1 Endoglucanase I

EG I is the major endoglucanase secreted by *T. reesei*, whose purification has been reported by several authors [46, 68, 74, 78, 79, 86–93]. The precise structure of EG I has been revealed by the isolation of its gene [94, 95]. It codes for a 437 aa long polypeptide with an M_r of 46 kDa. The corresponding aa composition coincides very well with that of EG I proteins purified by Shoemaker et al. [68] and Bhikhabhai et al. [79]. It also coincides with aa compositions of endoglucanases purified by other workers (see Table 4), if their molecular weight determinations are considered erroneous (cf. 4.1.1) and adjusted to 55 kDa. The native M_r of EG I, when determined by LDSM is 52 110 Da, indicating that a M_r of 6.1 Da is due to carbohydrate residues. The nature of this carbohydrate residues is however unclear: according to the aa sequence, 6 putative N-glycosylation sites would be present (at aa 78, 164, 204, 208, 281, and 394). However, PDMS of EG I CNBr-fragments showed that those aa at 78 and 164 are obviously not occupied. Nevertheless, the size of the glycosylated fragments was not compatible with the usual $(Man)_9(GlcNAc)_2$ or $(Man)_5(GlcNAc)_2$ structure [20]. Salovuori [96] reported that 70% of EG I carbohydrate was O-glycosidically linked. This would indicate that less of the resulting N-glycosylation sites are occupied in EG I, and that the differences in M_r found in the corresponding fragments must be due to other reasons (i.e., adsorption of cellooligosaccharides, non-enzymatic glucosylation etc.). Postsecretional modification appears to contribute only little to the strikingly high number of multiple forms of endoglucanases found by some workers [30–32, 37, 51, 52], since the enzyme is comparably resistant to proteolysis [46] and enzymatic deglycosylation [28]. The purification of a 50 kDa endoglucanase I, which can be identified according to its aa composition, and lack of detectable glycosylation [89] as EG I is so far the only result pointing to a possible postsecretional modification. Stahlberg et al. [97] cited unpublished results to have purified a truncated, AB-region less EG I from *T. reesei* culture filtrates. The occurrence of such a fragment has indeed been observed in some commercial cellulase preparations, albeit in very low amounts [98]. Since EG I exhibits a strong tendency for di- and trimerization, these oligomers are very stable and SDS-resistant, and can only be separated by treatment with non-polar solvents, indicating the involvement of hydrophobic forces in the aggregation process [99]. The EG I displays di- and trimer bands at IEP 4.7 and 5.1 (unpublished data), hence being responsible for at least two of the multiple EG I bands seen in IEF or chromatofocusing [100].

Table 4. Properties of various endoglucanases from *T. reesei*

Ref.	EG I						EG III				"small EG"		"EG III"
	[79]	[59]	[78]	[89]	[91]	[91]	[78]	[101]	[101]	[91]	[105]	[103]	[79]
Ala	5.5	5.9	6.0	5.8	5.7	7.3	7.2	7.0	7.8	8.0	8.6	8.6	11.6
Arg	1.8	1.8	2.0	1.7	1.8	1.4	2.6	2.5	2.8	2.8	1.2	2.0	1.9
Asx	13.9	12.7	13.6	13.8	12.8	12.7	13.7	12.3	13.1	12.3	10.5	11.1	7.0
Cys	5.0	4.7	4.1	ND	5.1	3.3	ND	3.0	ND	3.0	2.9	2.0	3.6
Glx	6.6	7.0	7.0	6.8	6.8	6.6	7.5	8.0	7.5	8.5	11.5	7.6	7.0
Gly	11.1	11.0	11.7	12.4	11.1	10.7	11.6	10.6	12.0	11.8	16.3	14.7	10.0
His	1.4	2.0	1.5	1.0	ND	1.2	1.1	1.2	1.4	1.5	6.4	1.5	0.7
Ile	2.7	2.9	2.8	3.1	2.8	3.8	4.5	5.3	5.0	4.8	3.0	3.3	3.6
Leu	5.5	5.4	6.0	6.2	5.4	5.7	6.0	5.8	6.1	5.2	4.9	4.0	3.9
Lys	2.3	2.1	2.3	2.4	2.2	1.6	1.5	1.5	2.1	1.7	4.6	2.5	2.0
Met	2.0	2.2	2.0	2.4	2.2	1.2	0.8	1.0	0.9	0.7	1.3	1.0	0.4
Phe	2.0	2.2	2.2	2.4	2.0	2.1	3.8	3.2	3.3	3.0	2.5	2.5	1.8
Pro	5.5	5.6	5.7	5.5	5.3	7.6	6.0	4.7	5.2	5.8	ND	4.6	6.7
Ser	13.4	13.5	12.9	11.7	13.4	11.8	8.7	10.5	11.7	9.6	8.5	11.7	11.7
Thr	10.5	10.3	10.6	10.3	10.3	11.1	10.9	11.0	11.4	10.0	5.8	9.1	15.0
Tyr	5.2	5.3	5.4	6.2	5.1	4.0	3.3	3.5	3.6	3.3	ND	5.5	9.1
Trp	1.6	1.6	ND	2.4	1.4	2.6	3.0	2.8	ND	2.5	2.1	2.5	1.6
Val	4.3	4.5	4.9	4.2	4.3	5.2	6.0	5.5	5.9	5.3	6.6	2.8	7.1
[kDA]	46	55	54	44.7	51	52	34.5	42.2	48.0		25	20.5	48
CH [%]	ND	11	ND	12	0	ND	15	18	15		0	0	12

Abbreviations as in Table 3

4.2.2 Endoglucanase III

A second EG, characterized in remarkable detail, is EG III. Three subspecies of this enzyme have been purified [97, 101], which display M_r figures of 48, 48 and 37 kDa, and corresponding IEPs of 5.7, 5.1 and 4.8. The 48/5.7 and 37/4.8 proteins exhibited amino acid compositions closely resembling those reported for EG IV and II [91]. The amino acid sequence, as deducible from the genes' sequence' confirms the usual core-AB structure and the corresponding amino acid composition confirms the 48/5.7 (EG III$_1$) isoenzyme as the primary gene product. It contains 47 mannose residues, which all — with the exception of a single Man$_9$-antenna at aa 103 — are located within the B-region. Partial amino acid sequencing of the 37/4.8 (EG III$_2$) protein revealed that it lacked the first 61 N-terminal amino acids, while otherwise being identical to EG III1 [97]. Consequently, only 9 mannose residues are present in this truncated protein.

The nature of the third EG III (48/5.1) has not yet been identified. However, CNBr-peptide mapping, provided a very similar fingerprint to that of EG III1 suggesting that it is a very closely related subspecies. It should be noted, that this protein was still accompanied by a second band of about 40 kDa in the final preparation. A complex between EG III$_1$ and this other protein (whose nature had not been identified) may explain the occurrence of this third EG III peak.

4.2.3 Other Endoglucanases

A number of other endoglucanases have been purified from *T. reesei* culture filtrates, which cannot be identified as EG I or EG III, and therefore, deserve special comments: of particular interest is the 43 kDa/IEP 4.0 endoglucanase purified by Niku-Paavola et al. [92]. Due to the lack of data on its aa composition, it is impossible to align this enzyme to EG I or III. However, its low carbohydrate content (2–7%) suggests that it constitutes a proteolytic fragment. The observed modification of this protein to a higher M_r form upon storage may be due to further proteolysis and/or aggregation. In view of the reputed high specific activity of endoglucanase, a knowledge of more details on its molecular characteristics would be worthwhile.

Pettersons' group have purified a 48 kDa/IEP 4.5 endoglucanase [79], which, however, differs from EG III1 with respect to its amino acid composition and N-terminal sequence (PyrGlu-Thr-Arg vs. PyrGlu-Glu-Thr-Val). Its amino acid composition and carbohydrate content is sufficiently unique to postulate that this protein is not related to either EG I or III and therefore, should be the product of a further EG gene, whose isolation, however, has not been attempted so far.

Some authors [74, 102–105] have also purified a "low-molecular weight" endoglucanase (20, 23.5 and 25 kDa, resp., IEP 7.2–7.5), which lacks glycosylation. The nature of this protein is still unclear. Its formation appears to depend strictly on the presence of cellulose in the medium, unlike all other cellulases, which are also formed in the presence of lactose or sophorose [99, 106]. Furthermore, it was shown to react with a polyclonal antibody against EG I [99]. In the EG I gene, the first N-terminal intron contains a TGA stop codon. Loss of splicing and therefore, termination at this side could result in the formation of a 27 kDa protein lacking most of EG I's glycosylation sites. However, whereas the amino acid composition of the 20.5 kDa EG [103] corresponds with that of this putative protein in 13 of 17 aa, the aa content reported by another group [105], and the N-terminal aa sequence, is clearly different. This contradicts the origin of this enzyme from differential splicing, and also makes its origin from partial proteolysis unlikely. These conclusions are supported by the lack of reactivity of monoclonal antibodies against EG I, which detect various EG I degradation products, with endoglucanase of "low-molecular weight" [46]. Further investigations on this enzyme are clearly needed but are complicated by the enzymes lack of glycosylation. The enzyme is rapidly degraded by the fungus' extracellular proteases (unpublished results), and − at best− constitutes only a very small portion of the secreted proteins. The apparent lack of glycosylation would also mean that this enzyme should lack the usual AB-motives, which would render this cellulase an interesting topic for further studies.

4.3 β-Glucosidases

Our knowledge on *Trichoderma* β-glucosidase is still poor in comparison to cellobiohydrolases and endoglucanases, although its purification has been reported by several authors [47, 68, 78, 107–114]. This is in part due to the fact that its gene has only very recently been isolated [115], and its sequencing is still being investigated. Moreover, β-glucosidase represents only a very low portion (0.5–1%) of the total extracellular protein mixture secreted by *T. reesei*, and considerable difficulties are observed in isolation of an intact, homogenously purified preparation [116]. The properties of different preparations obtained are given in Table 5: although little coincidence is deducable from this compendium, some evidence emerges that a 70–80 kDa protein of alkaline isoelectric point (8–8.4) is frequently found. Chirico and Brown [110] report that this enzyme contains about 50% hydrophobic amino acids, and a single N-linked sugar antenna. Hofer et al. [47], using monoclonal antibodies, reassessed the origin of multiple forms of *T. reesei* β-glucosidase. They provided evidence that partial proteolysis of the 80 kDa, IP 8.4 β-glucosidase results in the formation of 50 and 35 kDa protein fragments, which are still active and exhibit isoelectric points of 6.1 and 5.7, respectively. It is thus very likely that some of the β-glucosidases, described in [107, 108, 113] are proteolytic fragments of the 80 kDa β-glucosidase. This assumption is strengthened by the fact that several commercial cellulase preparations of *T. reesei*, contain high protease activity, and lack an intact β-glucosidase protein [98]. However, the occurrence of a 100–120 kDa β-glucosidase, observed by some authors [68, 98] cannot be explained in this way. It is possible that this enzyme is identical to the intracellular β-glucosidase of *T. reesei* [117]. This enzyme does not react with monoclonal antibodies against the 80 kDa β-glucosidase [47]. The role and intracellular location of this enzyme is still unclear, but it has been proposed to function in the controlling of the accumulation of cellulase inducers [118]. It is possible that this enzyme can be secreted in certain strains or under certain conditions; alternatively, its extracellular occurrence may be due to autolysis.

Table 5. Properties of β-glucosidases isolated from *T. reesei*

Ref.	[78]	[107]	[78]	[110]
Asx	11.8	12.5	13.4	12.5
Thr	7.4	7.1	7.8	6.8
Ser	7.4	8.2	9.5	9.6
Glx	6.2	6.7	7.2	6.9
Pro	9.0	5.6	7.0	5.0
Gly	11.0	10.8	11.4	10.5
Ala	10.3	10.1	10.3	10.0
Cys	ND	1.3	0.6	0.8
Val	8.8	8.2	5.1	9.0
Met	0.2	1.3	1.0	1.0
Ile	4.8	4.8	4.3	4.3
Leu	7.0	7.1	6.6	7.1
Tyr	3.6	3.5	3.7	3.4
Phe	2.8	2.6	3.9	2.5
His	1.4	1.2	1.4	1.2
Lys	4.0	3.0	2.9	3.3
Arg	3.8	3.4	3.3	3.0
Trp	ND	21	ND	3.0
[kDa]	75.6	47.0	76.6	79.7
CH [%]	0	0	0	1.3

Symbols used as in Table 3

The low amount of β-glucosidase in the culture fluid has been shown to be due to a predominant amount bound to the cell wall [119]. The cell-wall bound enzyme has been solubilized by treatment with *Aspergillus niger* glycosidases, and characterized, showing that the cell-wall bound β-glucosidase activity is due to a single, specific enzyme, resembling the 80 kDa enzyme with respect to size, charge and reactivity against monoclonal antibodies [120, 121]. The cell wall binding has previously been attributed to association with β-1,3-glucan [122–124]; however, more recent studies showed that upon enzymatic removal of the β-1,3-glucan, a β-glucosidase-heteroglycan complex is released, which can be separated by anion exchange chromatography [125]. The heteroglycan, whose structure has been analyzed recently [48] (Fig. 4), is a constitutive component of the *T. reesei* cell

```
–α-D-Manp-(1→6)-α-D-Manp-(1→6)-α-D-Manp-(1→6)-α-D-Manp-(1→6)–
         |                                    |
   α-D-Manp(1→2)                        β-D-Galf(1→2)
                                              |
                                        β-D-Galf(1→6)
                                              |
                                        α-D-GlcAp(1→2)
                                              |
                                        [α-D-Glcp(1→4)]
```

Fig. 4. Chemical structure of the heteroglycan secreted by *T. reesei*. The bracket indicates a glucose residue which is a regular constituent of the cell-wall bound glycan, which occurs only in a lower portion of the extracellularly found glycan

wall, and forms a strong complex with β-glucosidase upon incubation in vitro. Most interestingly, this heteroglycan activates β-glucosidase and other glycosidases, both from *Trichoderma* as well a other fungal species [48]. We have detected up to 0.3 mg mg^{-1} protein of the glycan in commercial cellulase preparations; hence, determination of β-glucosidase activities in crude cellulases may lead to over estimations of their β-glucosidase content.

   T. reesei also contains a plasma-membrane-bound β-glucosidase of 70 kDa and IP 8.4 [126], which reacts with monoclonal antibodies against the 80 kDa enzyme [47]. This enzyme probably plays a role in the mechanism of cellulase induction [127, 128]; however, it may also be an intermediate of the β-glucosidase secretory pathway.

5 Protease Activity and Proteolytic Events in *Trichoderma* Cellulases

The data shown before, provided a strong evidence for a proteolytic origin for some of the cellulases purified. The occurrence of proteases in *T. reesei* culture fluids has been reported by several authors [30, 38, 129, 130]. A recent screening of a number of commercial cellulase preparations demonstrated that proteolysis can be particularly strong in some preparations, and is especially pronounced with CBH I and CBH II, and to a lesser extent with β-glucosidase [98]. EG I appears to be less susceptible to proteolysis, eventually because of its occurrence as a dimer [46]. The protease(s) involved in cellulase degradation are specifically induced by an extracellular pH below 4 and by already low extracellular protein concentrations (around 0.1 mg ml^{-1}) [131]. This specific requirement for a low pH may explain why some authors, working with culture conditions which kept the pH constant [30, 31] reported an absence of proteolysis. The main protease component has been purified [131] and shown to belong to a class of pepsatin-insensitive aspartate protease [132]. Different cellulases appear to become attacked in a different way by this enzyme: degradation of CBH I exclusively starts by removing the AB-domain, whereas CBH II becomes attacked simultaneously from both termini [39]. Once the termini are removed, degradation commences considerably, hence, providing some proof that the termini somehow protect the enzymes from proteolysis. Since the termini also contain the cellulose adsorbing domain, proteolytically attacked cellulases fail to adsorb (and hence attack) crystalline cellulose [133, 134]. Proteolysis may therefore adversely effect the quality of cellulase preparations. Cloning of the protease genes, followed by their subsequent inactivation or disruption is an option for improvement.

6 Cellulase Enzyme Reactions and the Degradation of Cellulose

Although it has been illustrated above that there is already some detailed knowledge available on the structure and molecular properties of the main cellulase components of *T. reesei*, a major problem is still encountered when it comes to the understanding of how cellulases bring about the degradation of crystalline cellulose. This area is being strongly investigated at the moment, but still no conclusive model can be presented.

The postulation of a model on cellulose degradation obviously depends on firm knowledge of the kinetics and substrate specificities of individual cellulases. Investigations in this area have been crowned by success, especially because of

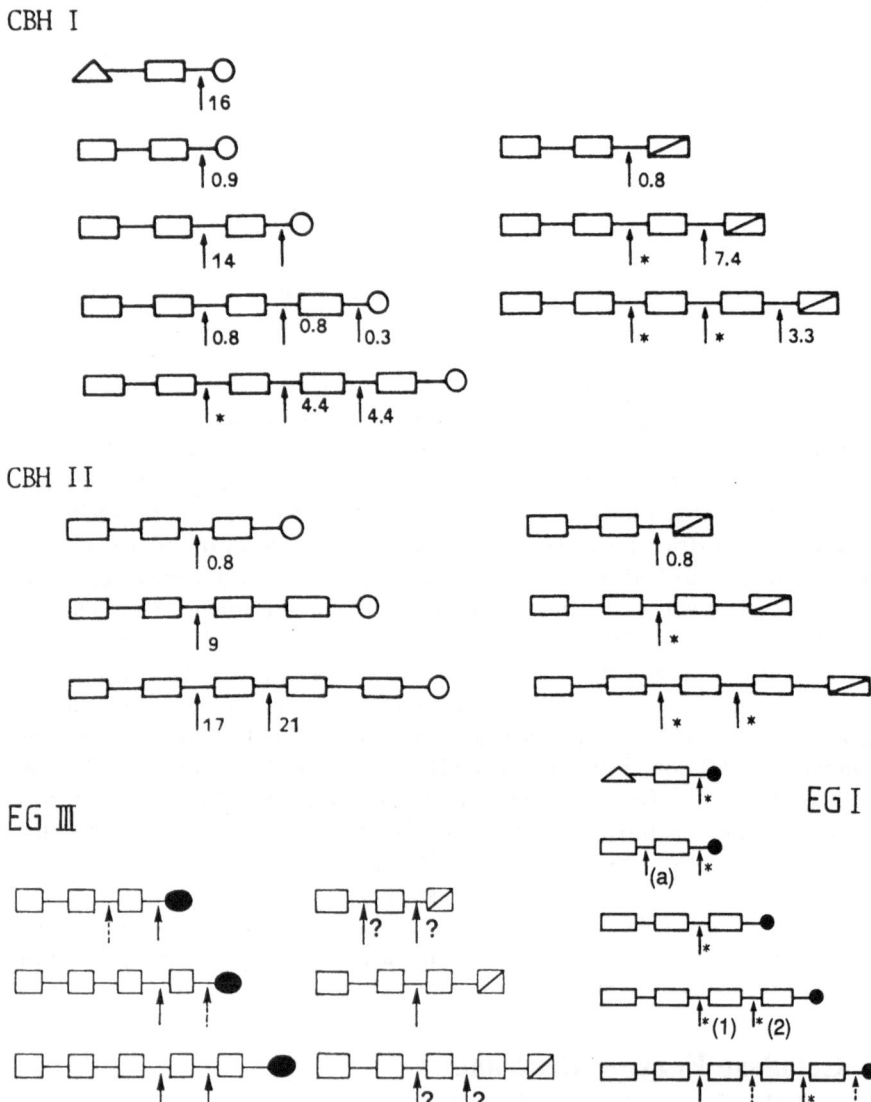

Fig. 5. Substrate specificities of CBH I, CBH II, EG I and EG III from *T. reesei*, using low molecular weight derivatives of the cellooligosaccharides. Symbols: □, 1,4-D-glucopyranosyl; △, 1,4-D-galactopyranosyl; ▨, reducing end; ○, ●, 4-methyl-umbelliferyl-; the arrow indicates the hydrolysis site, and the number specifies the turn over number (min⁻¹); * not determined. Taken from Refs. [101, 149, 200], by permission

the introduction of a low molecular mass chromogenic 4-methylumbelliferyl-β-glycosides [135], and by the expression of cloned cellulase genes in *Saccharomyces cerevisiae*, which eliminated the problem of cross-contamination of enzymes purified to homogeneity [95, 136, 137]. A comparison of the results obtained is given in Fig. 5. They show pronounced differences in the kinetics of glycoside hydrolysis, not only between the CBHs and EGs, but also between CBH I and II, and EG I and EG III. In the case of the cellobiohydrolases and the hydrolysis of cellooligosaccharides (d.p. < 3), little differences are apparent between the two enzymes. However, with higher homologues differences occur. CBH II shows a more strict substrate specificity, three to four contiguous β-1,4-linked glycosyl residues being required. The apparent K_m values are almost constant and turn-over numbers increase steadily as a function of chain length, more typical for an endo-type enzyme [138–140]. CBH I, hydrolyzing both cellobiosides and lactosides, attacks at sites other than the non-reducing end usually the cellobiosyl residues in the higher homologues [141]. The influence of chain length on the kinetic parameters shows no definite trend. These findings have been critisized as not being in accordance with its classification as an "exo-cellobiohydrolase" [144–146], and its classification as 1,4-β-D-glucan cellobiohydrolase has thus been recommended [148]. Moreover, CBH I is acting with overall retention of configuration, but CBH II with overall reversion [142, 143]. Hence, catalysis by CBH I apparently operates via an enzyme-glycosyl-intermediate, whereas CBH II carries out a single displacement by a nucleophilic water molecule.

The reactions catalyzed by the endoglucanases have for a long time been disputed, mainly due to the difficulties encountered in purifying these enzymes to homogeneity. It became clear now that *T. reesei* contains "non-specific" as well as "specific" EGs [147]. EG I clearly belongs to the group of non-specific EGs since it hydrolyzes not only cellulose, but also xylan [37, 147]. Its apparent hydrolytic parameters (k_{cat}, k_{cat}/k_m) increase steadily as a function of the number of glucose residues. An important feature of its hydrolytic reaction is that it proceeds via transfer reactions [149] (Scheme 1). This indicates that EG I probably possesses several subsites, each capable of binding a glucose molecule; during the hydrolysis reaction, glycosyl intermediates are formed and these can be transferred to acceptor molecules. EG acts via retention of its configuration [149]. In view of this data, Biely [147] recommends to term this enzyme as 1,4-β-D-glucan-1,4-β-xylan-glucanoxylanohydrolase, and postulates the definition of a new EC code.

T. reesei also secretes two "specific" EGs, EG III and the less well defined "low MW-EG" [147]. No details on their reaction mechanism are available yet, however.

(1)	$E + MeUmb(Glc)_2 \longrightarrow E(Glc)_2 + MeUmb$	Hydrolysis
(2)	$E(Glc)_2 + MeUmb(Glc)_2 \longrightarrow E + MeUmb(Glc)_4$	Self-transfer
(3)	$E + MeUmb(Glc)_4 \longrightarrow E + (Glc)_3 + MeUmbGlc$	Hydrolysis
(4)	$E + MeUmb(Glc)_4 \longrightarrow E + (Glc)_2 + MeUmb(Glc)_2$	Hydrolysis
(5)	$E + (Glc)_3 \longrightarrow E + (Glc)_2 + Glc$	Hydrolysis

Scheme 1

The enzymatic hydrolysis of a glycosidic bond is commonly assumed to proceed with a lysozyme-type mechanism through protonation of the glycosidic oxygen by an acidic amino acid residue [150, 151] and stabilization of the resulting carbonium ion by another amino acid residue. The active center of CBH I and EG I have been studied by looking at the effects of chemical modifications [152, 153] and site-directed mutagenesis [154]. Evidence thereby obtained suggests that the glutamate residue at aa 126 (CBH I) and aa 127 (EG I), respectively, are involved either in substrate binding or stabilization.

As emphasized earlier, the enzymic degradation of cellulose has originally been interpreted by a synergism of an enzyme, rendering cellulose accessible and enzymes subsequently degrading the exposed oligosaccharidyl chains (C1-Cx-concept [7]). With a growing body of data on the properties of cellulases this concept has become modified throughout the last two decades. The major obstacle is the still missing evidence for the identity of the initial C1-enzyme. In view of the appearance of oxidized cellooligosaccharides during the course of cellulolysis by *T. reesei* [155], an oxidative enzyme has been postulated to participate in the initial attack on crystalline cellulose. Indirect evidence for the existence of such an enzyme had been presented [156]; however, even by an extensive screening of cofactors, we have been unable to detect the activity of such an enzyme by assay [157]. It should be noted that oxidized end groups are known to occur in cellulose [158], and hence may themselves provide targets for initial enzymic attack. On the other hand, a „microfibril generating factor" (MFG) was isolated from a commercial cellulase preparation [159, 160]. It appears to be a low molecular weight glycopeptide, whose activity is increased by the presence of ferric ions. The identity of this substance clearly challenges further studies. In any case, it is uncertain whether initial "amorphogenesis" of cellulose is at all necessary, since "microcracks", (i.e., disturbances in the crystalline structure of cellulose) are commonly observed in nature.

The number of cellulases involved in the degradation of cellulose is still speculative, but several types of synergistic degradation have been observed with celluloses of different degree of crystallinity [161]. Hence CBH I and CBH II synergistically cooperate in the degradation of crystalline cellulose, whereas synergism between other cellulases has been observed with substrates of intermediate crystallinity. This "exo-exo"-synergism of CBH I and II [162] has challenged interpretation: Wood [163] proposed that the two enzymes might display specificities against two different types of non-reducing end groups, one being exposed from, whereas the other being buried in the cellulose surface, as shown by [164]. A kinetic investigation of the adsorption of CBH I and II to cellulose showed that the binding of both CBH I and CBH II to cellulose is slow (equilibration time > 40 min), and (at least in the case of Avicel) almost irreversible (K_d $104 \, M^{-1}$), yielding a figure of 110 moles mg^{-1} of cellulose for both enzymes [165]. These authors also provided evidence for prior formation of a "loose" complex of CBH I and CBH II which is necessary for adsorption. This may indicate that the ability of various cellulases from *T. reesei* to form complexes with each other [42, 43], may indeed have a biological function.

Another interesting point is that the "natural" mixture of cellulolytic enzymes secreted by *T. reesei* is not identical to the one giving optimal synergism [161]; in fact, CBH I is far in excess. In view of the recent development of convenient gene transformation systems for *T. reesei* [166–170], the secretion of "tailor-made" enzyme mixtures by recombinant strains containing either, amplified copies of particular cellulase genes or expressing them under a stronger promotor was acchieved [171–173].

7 Regulation of Cellulase Formation

This subject has been reviewed recently and the reader must therefore, refer to the following article for a detailed survey [174]. However, some complimentary comments should be made: as an insoluble substrate, cellulose cannot penetrate the cell-wall and cell-membrane, and hence can not be the inducing component. This appears to be logical in view of the differences in the structure of different cellulosic substrates in nature. Therefore, *T. reesei* and other *Trichoderma* species contain cellulases located on the surface of the conidia, which appear to be involved in the initial attack on cellulose [127, 175]. An interesting feature of the conidial bound cellulase spectrum is that here CBH II is the major conidial-bound cellulase, and that in some higher producer strains, higher amounts of conidial CBH II are present [176]. The reason for this preferential synthesis of CBH II during conidiation (CBH II makes up only a minor portion of extracellular cellulase protein in mycelial cultures, see Table 1) still deserves investigation; of particular interest is the report by Chen et al. [85] of a putative cyclic AMP-dependent transcription activator binding region about 400 bp upstream of the start point of the *cbh 2* gene; a correlation between cyclic AMP levels and sporulation in *T. viride* has been reported [177].

These "initial degradation products" can be either taken up by the mycelium or be hydrolyzed by the cell-wall- and plasma-membrane bound β-glucosidase [64]. Also, the formation of transglycosidation products by β-glucosidase [178] and EG I [148, 149] has been reported. The nature of the inducing oligosaccharide is, however, still unknown: two compounds − sophorose and cellobiono-δ-1,5-lactone have been shown to be very potent inducers of cellulase formation in resting mycelia [179–181]. Both components have also been shown to occur in *T. reesei* culture filtrates during growth on cellulose [155, 182], and the introduction of additional oxidized end-groups into cellulose (which are supposed to release higher amounts of cellobiono-δ-1,5-lactone during hydrolysis) increases the rate of cellulase formation [183]. However, it is still unclear whether any of these components are involved in cellulase induction by cellulose under normal conditions. Both inducers are characterized by the fact that they are poor substrates for *T. reesei* β-glucosidase [157], and therefore, guaranteeing that − in comparison to cellobiose − a smaller portion of these compounds are hydrolyzed whereas, a higher portion can enter the cell of the fungus. Proof for this assumption has recently been obtained by the finding that cellobiose can induce cellulase formation as efficiently as sophorose when: (a) β-glucosidase activity is simultaneously inhibited by specific inhibitors [64], or (b) when a β-glucosidase defective strain is used [128]. A comparison of β-glucosidase and β-linked disaccharide transport of *T. reesei* revealed that the V_{max} and K_m (substrate: cellobiose) are about 100–300-fold lower for permease than β-glucosidase [64]. Hence, at low concentrations of cellobiose, uptake is strongly favoured. Since the steady-state concentration of cellobiose is low during growth on cellulose, cellobiose may well be the „true inducer".

It appears at this point that the potential of physiological strategies to unravel this problem has now been exhausted; with the development of genetic techniques for *T. reesei* [166–170], further progress can only be expected by using molecular biological techniques. Only preliminary data is so far available in this area: three groups have recently reported [65, 184, 185] evidence suggesting that regulation of cellulase formation occurs at the level of transcription. Interestingly, strains accumulating higher cellulase amounts in the medium display higher steady-state levels of CBH I- and CBH II-mRNA in the mycelia, suggesting that transcription may limit cellulase production in lower producer strains. [186]. Also, the introduction of up to 20 copies of the *cbh 2* gene into the genome of *T. reesei* leads only to a 3–4-fold increase in CBH II production [173]. Band-shift assays [187], using protein from cell-free extracts of *T. reesei* and 5' upstream sequences of *cbh 1* and *cbh 2* indicate the presence of only low amounts of proteins binding to these sequences [186]. All this evidence suggests that cellulase formation may be limited at the level of transcription. Hence, the mechanism of activation of cellulase gene transcription is certainly the next challenge in our understanding of how cellulase formation is regulated.

Cellulase formation is generally considered to be subject to carbon catabolite repression [174, 188]. Proof for this claim is taken from the fact that almost no cellulase is formed during growth of *T. reesei* on glucose, glycerol and other carbon sources related to glycolytic metabolism [188]. However, we have recently shown that the addition of glucose to *T. reesei* cultures producing cellulases only decreases the rate of formation of CBH I (measured immunologically) to about 40% and continues to produce cellulases for up to 40 h [65]. A similar result was obtained when a number of glucose analogues or inhibitors of glucose uptake or metabolism were used; finally, the rate of cellulase production was similar when the mycelia were transferred to a medium lacking a new carbon source. Such a finding is not compatible with a control of cellulase formation by catabolite repression. On the other hand, several authors reported on the isolation of mutants of *T. reesei*, which were "carbon catabolite derepressed" [189–192]. It is questionable whether this term is in fact applicable to these mutants: in all cases they did not form cellulases upon growth on glucose, but only produced cellulases in the presence of some glucose levels in addition to an inducing carbon source. There is no evidence that glucose or a catabolite thereof in fact controls the transcription of cellulase genes, as would be suggested by the term "catabolite repression", and it is therefore recommended not to use this term. The only alteration detected in the respective mutants is at the level of the endoplasmic reticulum [193–195], suggesting that glucose may effect the development of the secretory pathway or some of its components. O-Glycosylation, which takes place at the endoplasmic reticulum

[25], has been shown to be a step controlling the secretion of cellulases by *Trichoderma* [27, 28]. A decreased activity of dolichol-phosphate-mannose synthase, a key enzyme in the O-Glycosylation pathway in *T. reesei*, has been found during growth on glucose [25]. While it remains to be proven whether the action of glucose on cellulase formation is actually due to an effect on secretion, it becomes clear that the role of glucose in cellulase formation needs still careful analysis.

8 Conclusions

More than 40 years after the discovery of the cellulolytic potential of *Trichoderma* (for a survey of the historical dimension see Ref. [196], the secret why this organism produces cellulases so efficiently has only in part been unraveled. Research in this area certainly is curtailed by the fact that selected strains of *T. reesei* already produce extracellular cellulases in excess of 30 g l^{-1} [197], a yield which is unlikely to be increased further. A more detailed knowledge in this area may however, be important in order to increase the production of other enzymes by this fungus. *T. reesei* has been recommended as a powerful recombinant producer of mammalian secretory proteins, i.e. chymosine or vaccines [172], yet the yields obtained today are still disappointingly low. Moreover, even homologous proteins of *T. reesei* are produced only at a comparably low level under the control of the promotor and terminator sequences of *cbh 1*, which is secreted in amounts of 25–30 g l^{-1} [171]. Hence, other factors involved in the formation and secretion of cellulases by this fungus are apparently of great importance. With the increasing interest in the high-yield production of other *Trichoderma* enzymes, i.e., xylanases [198], this subject will deserve growing interest in future.

Probably the most exciting aspect of cellulase biochemistry currently dealt with is the discovery of how the three-dimensional structure of cellulases manages the degradation of cellulose. These investigations may eventually soon offer prospects for the improvements of cellulases by modifying their active center or the structures involved. They may also enable a scientific explanation for the requirement of different ratios of individual cellulase components for the optimal hydrolysis of different cellulose sources. In this regard, we should be aware of the fact that we probably have not yet sufficiently identified all enzymes truely involved in cellulolysis by this fungus.

Finally it should be noted that there is still very little molecular biological knowledge avialable about how the presence of cellulose triggers cellulase biosynthesis in *T. reesei*. While this may be considered a merely academic exercise, it may nevertheless, yield important information about how a message is transmitted in an industrially important fungus.

Acknowledgements: The author is indebted to all those persons, who sent him papers „in press" or "submitted" as privileged information. He also gratefully acknowledges support of his own research in this area throughout the past 6 years by Fond zur Förderung Wissenschaftlicher Forschung, Bundesministerium für Wissenschaft und Forschung, Jubiläumsfond der Nationalbank und Hochschuljubiläumsstiftung der Stadt Wien.

9 References

1. Parisi F (1989) Advances in lignocellulosics hydrolysis and in the utilization of the hydrolyzates. In: Fiechter A (ed) Springer, Berlin, Heidelberg, New York p 53 (Advances in biochemical engineering/biotechnology, vol 38)
2. Mandels M (1982) Annu Rep Ferm Proc 5: 35
3. Pokorny M, Zupancic, Steiner T, Kreiner W (1990) Production and down-stream processing of cellulases on a pilot plant scale. In: Kubicek CP, Eveleigh DE, Esterbauer H, Steiner W, Kubicek-Pranz EM (eds) Trichoderma cellulases: Biochemistry, Physiology, Genetics and Application. Royal Chemical Society, Cambridge p 168
4. Enari TM (1983) Microbial cellulases. In: Fogarty WM (ed) Microbial enzymes and biotechnology. Applied Science London p 183
5. Mandels M (1985) Biochem Soc Trans 13: 414
6. Brown RM Jr (ed) (1982) Cellulose and Natural Polymer Systems Biogenesis, Structure and Degradation. Plenum Press New York
7. Mandels M, Reese ET (1964) Dev Ind Microbiol 5: 5
8. Beguin P (1990) Annu Rev Microbiol 44: 219
9. Van Tilbeurgh H, Tomme P, Claeyssens M, Bhikhabhai R, Petterson LG (1986) FEBS Letts 204: 223
10. Tomme P, Van Tilbeurgh H, Petterson LG, Vandekerckhove J, Knowles J, Teeri T, Claeyssens M (1988) Eur J Biochem 170: 575
11. Knowles JKC, Teeri TT, Lehtovaara P, Penttilä M, Saloheimo M (1988) The use of gene technology to investigate fungal cellulolytic enzymes. In: Aubert JP, Beguin P, Millet J (eds) Biochemistry and Genetics of Cellulose Degradation. FEMS Symp 43. Academic Press New York p 153
12. Schmuck M, Pilz I, Hayn M, Esterbauer H (1986) Biotechnol Letts 8: 397
13. Abuja PM, Schmuck M, Pilz I, Tomme P, Claeyssens M, Esterbauer H (1988) Eur J Biophys 15: 339
14. Abuja PM, Pilz I, Claeyssens M, Tomme P (1988) Biochem Biophys Res Comm 156: 180
15. Bergfors T, Rouvinen J, Lehtovaara P, Caldentey X, Tomme P, Claeyssens M, Petterson LG, Teeri T, Knowles JKC, Jones TA (1989) J Mol Biol 209: 167
16. Johansson G, Stahlberg J, Lindeberg G, Engstrom A, Petterson LG (1989) FEBS Letts 243: 389
17. Kraulis PJ, Clore GM, Nilges M, Jones TA, Petterson G, Knowles JKC, Gronenborn AM (1989) Biochem 28: 7241
18. Gardner H, Blackwell J (1974) Biopolym 13: 1975
19. Olden K, Bernard B, Humphries MJ; Yeo T-K, Yeo K-T, White SL; Newton SA, Bauer HC, Parent JB (1985) Trends Biochem 10: 78
20. Salovuori I, Makarow M, Rauvala H, Knowles JKC, Kääriänen L (1987) Bio/Technol 5: 152
21. Gum EK, Brown RD Jr. (1977) Biochim Biophys Acta 492: 225
22. Allmaier G, Hagspiel K, Kubicek CP (1991) manuscript in preparation
23. Gum EK, Brown RD Jr (1976) Biochim Biophys Acta 446: 371
24. Chasteigner du Mee CPR (1984) Thesis, University of Florida
25. Kruszewska J, Messner R, Kubicek CP, Palamarczyk G (1989) J Gen Microbiol 135: 301
26. Perera IK, Uzcategui E, Hakansson P, Brinkhalm G, Petterson G, Johansson G, Sundqvist BUR (1990) Rapid Comm Mass Spectrom 4: 285
27. Messner R, Kubicek CP (1988) FEMS Microbiol Letts 50: 227
28. Kubicek CP, Panda T, Schreferl-Kunar G, Gruber F, Messner R (1987) Can J Microbiol 33: 698
29. Shoemaker SP, Raymond JC, Bruner R et al (1981) Cellulases: Diversity among improved Trichoderma strains. In: Hollaender A (ed) Trends in the Biology of Fermentation for Fuels and Chemicals. Plenum Press New York p 89

30. Sheir-Neiss G, Montenecourt BS (1984) Appl Microbiol Biotechnol 20: 46
31. Kammel WP, Kubicek CP (1985) J Appl Biochem 7: 138
32. Farkas V, Jalanko A, Kolarova N (1982) Biochim Biophys Acta 706: 105
33. Beguin P (1983) Anal Biochem 131: 333
34. Biely P (1987) Differentiation of glycanases of microbial cellulolytic systems using chromogenic and fluorogenic substrates. In: Chaloupka J, Krumphanzl V, (eds) Extra-cellular enzymes of Microorganisms. Plenum Press New York. p 187
35. Van Tilbeurgh H, Claeyssens M, De Bruyne CK (1982) FEBS Letts 149: 152
36. Labudova I, Farkas V (1983) Biochim Biophys. Acta 744: 135
37. Biely P, Markovic O (1988) Biotechnol Appl Biochem 10: 99
38. Nakayama M, Tomita Y, Suzuki H, Nisizawa K (1976) J Biochem 79: 955
39. Hagspiel K, Haab D, Kubicek CP (1989) Appl Microbiol Biotechnol 32: 61
40. unpublished data, cited in: Vanhanen S, Penttilä M, Lehtovaara P, Knowles JKC (1989) Curr Genet 15: 181
41. Allen RC, Saravis RC, Maurer HR (eds) (1984) Gel electrophoresis and isoelectric focusing of proteins: selected techniques. Walter de Gruyter. Berlin.
42. Sprey B, Lambert C (1983) FEMS Microbiol Letts 18: 217
43. Sprey B, Lambert C (1984) FEMS Microbiol Letts 23: 227
44. Sprey B (1987) FEMS Microbiol Letts 43: 25
45. Sprey B (1987) FEMS Microbiol Letts 48: 211
46. Luderer MEH, Hofer F, Hagspiel K, Allmaier G, Blaas D, Kubicek CP (1991) Biochim Biophys Acta 1076: 427
47. Hofer F, Weissinger E, Messner R, Mischak H, Meixner-Monori B, Visser J, Blaas D, Kubicek CP (1989) Biochim Biophys Acta 992: 298
48. Rath J, Messner R, Kosma P, Altmann F, März L, Kubicek CP (1991) manuscript submitted
49. Weber M, Foglietti MJ, Percheron F (1980) J Chromat 188: 377
50. Labudova I, Farkas V, Bauer S, Kolarova N, Branyik A (1981) Eur J Appl Microbiol Biotechnol 12: 16
51. Bissett FH (1979) J Chromat 178: 515
52. Hayn M, Esterbauer H (1985) J Chromat 329: 379
53. Fliess A, Schügerl K (1983) Eur J Appl Microbiol Biotechnol 17: 314
54. Witte K, Heitz HJ, Wartenberg A (1990) Acta Biotechnol 10: 41
55. Ellouz S, Durand H, Tiraby G (1987) J Chromat 396: 307
56. Nummi M, Niku-Paavola ML, Enari TM, Raunio V (1980) FEBS Letts 113: 164
57. Fägerstam LG, Petteron LG (1979) FEBS Letts 98: 363
58. Szakacs-Dobozi M, Halasz A (1986) J Chromat 365: 51
59. Shoemaker SP, Watt K, Tsikovsky G, Cox R (1983) Bio/Technol 1: 681
60. Riske F, Labudova I, Miller L, MacMillan JD, Eveleigh DE (1987) In: Moo-Young (ed) Biomass Conversion Technology: Principles and Practice. Pergamon Press Toronto p 167
61. Aho S, Paloheimo M (1990) Biochim. Biophys. Acta 1087: 137
62. Mischak H, Hofer F, Weissinger E, Messner R, Hayn M, Tomme P, Küchler E, Ester-bauer H, Claeyssens M, Kubicek CP (1989) Biochim Biophys Acta 990: 1
63. Nieves RA, Himmel ME, Todd RJ, Ellis RP (1990) Appl Envir Microbiol 56: 1103
64. Fritscher C, Messner R, Kubicek CP (1990) Exp Mycol 14: 405
65. Messner R, Kubicek CP (1991) Appl Envir Microbiol 57: 630
66. Kolbe J, Kubicek CP (1990) App Microbiol Biotechnol 34: 26
67. Riske FJ, Eveleigh DE, MacMillan JD (1990) Appl Envir Microbiol. 56: 3261
68. Shoemaker SP, Schweickart M, Ladner D, Gelfand S, Kwok K, Myambo K, Innis M (1983) Bio/Technol 1: 691
69. Teeri TT, Salovuori I, Knowles J (1983) Bio/Technol 1: 696
70. Cheng C, Tsukagoshi N, Udaka S (1990) Nucl Acid Res 18: 5559
71. Fägerstam LG, Petterson LG, Engström JA (1984) FEBS Letts 167: 309
72. Nummi M, Niku-Paavola ML, Lappalainen A, Enari TM, Raunio V (1983) Biochem J 215: 677

73. Berghem LER, Petterson LG (1973) Eur J Biochem 37: 21
74. Beldman G, Searle-Van Leeuwen MF, Rombouts FM, Voragen FGJ (1985) Eur J Biochem 146: 301
75. Gritzali M, Brown RD Jr (1979) Adv Chem Ser 181: 237
76. Okada G, Nisizawa K, Suzuki H (1986) J Biochem 63: 591
77. Hostomska Z, Mikes O (1984) Int J Pept Prot Res 23: 402
78. Odegaard BH, Anderson PC, Lovrien RE (1984) J Appl Biochem 6: 156
79. Bhikhabhai R, Johansson G, Petterson LG (1984) J Appl Biochem 6: 336
80. Berghem LER, Petterson LG, Axiö-Frederickson UB (1975) Eur J Biochem 53: 55
81. Alluvalde JL, Ellenrieder G (1984) Enzyme Micr Technol 6: 467
82. Niku-Paavola ML, Lappalainen A, Enari TM, Nummi M (1986) Biotech Appl Biochem 8: 449
83. Van Tilbeurgh H, Bhikhabhai R, Petterson LG, Claeyssens M (1984) FEBS Letts 169: 215
84. Teeri TT, Lehtovaara P, Kauppinen S, Salovuori I, Knowles JKC et al (1987) Gene 51: 43
85. Chen CM, Gritzali M, Stafford DW (1987) Bio/Technol 5: 274
86. Emert GH, Gum EK Jr, Lang JH, Liu TH, Brown RD Jr (1979) Adv Chem Ser 181: 79
87. Gong CS, Chen LF, Tsao GT (1979) Biotechnol Bioeng 21: 167
88. Bhikhabhai R, Petterson LG (1984) FEBS Letts. 167: 301
89. Hakansson U, Fägerstam LG, Petterson LG, Andersson L (1979) Biochem J 179: 141
90. Shoemaker SP, Brown RD Jr (1978) Biochim Biophys Acta 523: 133
91. Shoemaker SP, Brown RD Jr (1978) Biochim Biophys Acta 523: 147
92. Niku-Paavola ML, Lappalainen A, Enari TM, Nummi M (1985) Biochem J 231: 75
93. Nieves RA, Ellis RP, Himmel ME (1990) Appl Biochem Biotechnol 24: 397
94. Penttilä M, Lehtovaara P, Nevalainen H, Bhikhabhai R, Knowles JKC (1986) Gene 45: 253
95. Van Arsdell JN, Kwok S, Schweickart VL, Ladner MB, Gelfand DH, Innis MA (1987) Bio/Technol 5: 60
96. Salovuori I (1987) Thesis. Tech Res Ctr Finland Publ. No 37
97. Stahlberg J, Johannson G, Petterson LG (1988) Eur J Biochem 173: 179
98. Gsur A, Kubicek-Pranz EM, Hayn M, Kubicek CP (1991) Biotechnol Appl Biochem, manuscript in press
99. Messner R, Gruber F, Kubicek CP (1988) J Bacteriol 170: 3689
100. Hagspiel K (1990) Thesis Technical University of Vienna
101. Saloheimo M, Lehtovaara P, Penttilä M, Teeri TT, Stahlberg J, Johansson G, Petterson LG, Claeyssens M, Tomme P, Knowles JKC (1988) Gene 63: 11
102. Gong CS, Ladisch MR, Tsao GT (1979) Adv Chem Ser 181: 261
103. Hakansson U, Fägerstam LG, Petterson LG, Andersson L (1978) Biochim Biophys Acta 524: 385
104. Hong SW, Hah Y-C, Maeng P-J, Jeong C-S (1986) Enzyme Micr Technol 8: 227
105. Ülker A, Sprey B (1990) FEMS Microbiol Letts 69: 215
106. Hrmova M, Biely P, Vrsanska M (1986) Arch Microbiol 144: 307
107. Berghem LER, Petterson LG (1974) Eur J Biochem 46: 295
108. Enari TM, Niku-Paavola ML, Harjo L, Lappalainen A, Nummi M (1981) J Appl Biochem 3: 157
109. Gong CS, Ladisch MR, Tsao GT (1977) Biotechnol Bioengin 19: 959
110. Chirico WJ, Brown RD Jr (1987) Eur J Biochem 165: 333
111. Schmid G, Wandrey C (1987) Biotechnol Bioengin 30: 571
112. Jackson MA, Talburt DE (1988) Biotechnol Bioengin 32: 903
113. Wood TM, McCrae SI (1982) J Gen Microbiol 128: 2973
114. Wilhelm M, Sahm H (1986) Acta Biotech 6: 115
115. Mach RL, Strauß J, Gonzales R, Kubicek CP (1991) manuscript submitted
116. Hofer F (1988) Thesis. Technical University of Vienna
117. Inglin M, Feinberg BA, Loewenberg JR (1980) Biochem J 85: 515
118. Loewenberg JR (1984) Arch Microbiol 137: 53

119. Kubicek CP (1981) Eur J Appl Microbiol Biotechnol 13: 226
120. Messner R, Kubicek CP(1990) Can J Microbiol 36: 211
121. Messner R, Kubicek CP (1990) Enzyme Microb Technol 12: 685
122. Kubicek CP (1982) Arch Microbiol 132: 347
123. Kubicek CP (1983) Can J Microbiol 29: 163
124. Kubicek CP (1983) FEMS Microbiol Letts 20: 285
125. Messner R, Hagspiel K, Kubicek CP (1990) Arch Microbiol 154: 150
126. Umile C, Kubicek CP (1986) FEMS Microbiol Letts 34: 291
127. Kubicek CP (1987) J Gen Microbiol 133: 1481
128. Strauss J, Kubicek CP (1990) J Gen Microbiol 126: 1321
129. Dunne CP (1982) Relationship between extracellular proteases and the cellulase system of Trichoderma reesei. In: Chibata J, Fukui J, Wingard Jr LB (eds.) Enzyme Engineering Vol 6. Plenum Press New York p 355
130. Lovrien RE, Gusek T, Hart B (1985) J Appl Biochem 7: 285
131. Haab D, Hagspiel K, Szakmary K, Kubicek CP (1990) J Biotechnol 16: 187
132. North MJ (1982) Microbiol. Rev. 46: 308
133. Gsur A (1990) Thesis. Technical University of Vienna
134. Chen HC, Grethlein HE (1988) Biotechnol Letts 10: 913
135. Van Tilbeurgh H, Claeyssens M (1985) FEBS Letts 187: 282
136. Penttilä M, Andre L, Saloheimo M, Lehtovaara P, Knowles JKC (1987) Yeast 3: 75
137. Penttilä M, Andre L, Lehtovaara P, Bailey M, Teeri TT, Knowles JKC (1988) Gene 63: 103
138. Claeyssens M, Van Tilbeurgh H, Tomme P, Wood TM, McRae SI (1989) Biochem J 261: 819
139. Van Tilbeurgh H, Petterson LG, Bhikhabhai R, de Boech H, Claeyssens M (1985) Eur J Biochem 148: 239
140. Van Tilbeurgh H, Loontines FG, Engelborgs Y, Claeyssens M (1989) Eur J Biochem 184: 553
141. Claeyssens M, Biely P (1990) unpublished data
142. Knowles JKC, Lehtovaara P, Murray M, Sinnott M (1988) J Chem Soc Chem Comm 1988: 1401
143. Claeyssens M, Tomme P, Brewer CF, Hehre EJ (1990) FEBS Letts 263: 89
144. Okada G, Tanaka Y (1988) Agric Biol Chem 52: 2981
145. Schmid G, Wandrey C (1990) J Biotechnol 14: 393
146. Enari TM, Niku-Paavola ML (1987) CRC Crit Rev Biotechnol 5: 67
147. Biely P (1990) Artificial substrate for cellulolytic glycanases and their use for the differentiation of Trichoderma enzymes. In: Kubicek CP, Eveleigh DE, Esterbauer H, Sten W, Kubicek-Pranz EM (eds) Trichoderma Cellulases. Biochemistry, Genetics, Physiology and Applications. Royal Chem Soc Cambridge p 30
148. Beldman G, Voragen AGJ, Rombouts FM, Searle-Van Leeuwen MF, Pilnik W (1988) Biotechnol Bioengin 31: 160
149. Claeyssens M, Van Tilbeurgh H, Kamerling JP, Berg J, Vrsanska M, Biely P (1990) Biochem J 270: 251
150. Capon B (1969) Chem Rev 69: 407
151. Paice MG, Jurasek L (1979) Adv Chem Ser 181: 361
152. Tomme P, Claeyssens M (1989) FEBS Letts 243: 239
153. Claeyssens M, Tomme P (1990) Structure-function relationships of cellulolytic proteins from Trichoderma reesei. In: Kubicek CP, Eveleigh DE, Esterbauer H, Steiner W, Kubicek-Pranz EM (eds) Trichoderma Cellulases: Biochemistry, Genetics, Physiology and Applications. Royal Chemical Society Press Cambridge. p 1
154. Mitsuishi Y, Nitisinprasert S, Saloheimo M, Biese I, Reinikainen T, Claeyssens M, Keränen S, Knowles JKC, Teeri TT (1990) FEBS Letts 275: 135
155. Vaheri M (1982) J Appl Biochem 4: 356
156. Vaheri M (1983) J Appl Biochem 5: 66
157. Szakmary K, Wottawa A, Kubicek CP (1991) J Gen Microbiol, in press
158. Schmidt E (1936) Ber Dtsch Chem Ges. 69: 366

159. Griffin H, Dintzis FR, Krull L, Baker FL (1984) Biotechnol Bioengin 26: 296
160. Krull LH, Dintzis FR, Griffin HL, Baker FL (1988) Biotechnol Bioengin 31: 321
161. Henrissat B, Driguez H, Viet C, Schülein M (1985) Bio/Technol 3: 722
162. Fägerstam LG, Petterson LG (1980) FEBS Letts 119: 97
163. Wood TM (1985) Biochem Soc Trans 13: 407
164. Henrissat B, Vigny B, Buleon A, Perez S (1988) FEBS Letts 231: 177
165. Tomme P, Heriban V, Claeyssens M (1990) Biotechnol Letts 12: 525
166. Penttilä M, Nevalainen H, Rättö M, Salminen E, Knowles JKC (1987) Gene 61: 155
167. Gruber F, Visser J, Kubicek CP, De Graaff L (1990) Curr Genet 18: 71
168. Gruber F, Visser J, Kubicek CP, De Graaff L (1990) Curr Genet 18: 447
169. Smith JL, Bayliss FT, Ward M (1991) Curr Genet, 19: 27
170. Cheng C, Tsukagoshi N, Udaka S (1990) Curr Genet 18: 453
171. Harkki A, Mäntylä A, Penttilä M, Muttilainen S, Bühler R, Suominen P, Knowles JKC,
     Nevalainen H (1991) Enzyme Microb Technol 13: 227
172. Uusitalo JM, Nevalainen H, Harkki A, Knowles JKC, Penttilä M (1990) J Biotechnol
     17: 35
173. Kubicek-Pranz EM, Gruber F, Kubicek CP (1991) J Biotechnol 20: 83
174. Bisaria V, Mishra S (1989) CRC Crit Rev Biotechnol 9: 61
175. Kubicek CP, Mühlbauer G, Krotz M, John E, Kubicek-Pranz EM (1988) J Gen
     Microbiol 134: 1215
176. Messner R, Kubicek-Pranz EM, Gsur A, Kubicek CP (1991) Arch Microbiol 155: 601
177. Farkas V, Gresik M, Kolarova N, Suolova Z, Sestak S (1990) Biochemical and physio-
     logical changes during photoinduced conidiation and derepression of cellulase synthesis
     in Trichoderma. In: Kubicek CP, Eveleigh DE, Esterbauer H, Steiner W, Kubicek-
     Pranz EM (eds) Trichoderma Cellulases: Biochemistry, Genetics, Physiology and
     Applications. Royal Chem Soc Press Cambridge p 139
178. Vaheri M, Leisola M, Kaupinnen V (1979) Biotechnol Letts 1: 41
179. Sternberg D, Mandels GR (1979) J Bacteriol 139: 761
180. Bruchmann EE, Graf H, Saad AA, Schrenk D (1978) Chem Ztg 102: 154
181. Iyayi CB, Bruchmann EE, Kubicek CP (1989) Arch Microbiol 151: 326
182. Mandels M, Parrish FW, Reese ET (1962) J Bacteriol 83: 400
183. Kubicek-Pranz EM, Steiner M, Kubicek CP (1990) FEMS Microbiol Letts 68: 273
184. El-Gogary S, Leite A, Crivellaro O, Eveleigh DE, El Dorry H (1989) Proc Natl Acad
     Sci US 86: 6138
185. Jokinen A, Penttilä M (1991) unpublished results; cited in Ref 172
186. Morawetz R, Gruber F, Messner R, Kubicek CP (1991) Current Genetics, in press
187. Fried MG (1989) Electrophoresis 10: 366
188. Nisizawa T, Suzuki H, Nisizawa T (1972) J Biochem 71: 999
189. Mishra S, Gopalkrishnan KS, Ghose TK (1982) Biotechnol Bioengin 24: 251
190. Beja da Costa M, Van Uden N (1980) Biotechnol Bioengin 22: 2429
191. Montenecourt BS, Eveleigh DE (1979) Adv Chem Ser 181: 289
192. Labudova I, Farkas V (1983) FEMS Microbiol Letts 20: 211
193. Ghosh A, Al Rabiai S, Ghosh BK, Trimino-Vasquez H, Eveleigh DE (1982) Enzyme
     Micr Technol 4: 110
194. Siegler KM (1986) Thesis Lehigh University. Bethlehem, Pennsylvania
195. Glenn M, Ghosh A, Ghosh BK (1985) Appl Envir Microbiol 50: 1137
196. El Gogary S, Leite A, Crivellaro O, El Dorry H, Eveleigh DE (1990) Trichoderma reesei
     cellulase — from mutants to induction. In: Kubicek CP, Eveleigh DE, Esterbauer H,
     Steiner W, Kubicek-Pranz EM (eds) Trichoderma cellulases: Biochemistry, Physiology,
     Genetics and Applications. Royal Chem Soc Press Cambridge p 200
197. Pourquie J, Warzywoda M, Chevron F, Thery M, Lonchamp D, Vandecasteele JP
     (1988) Scale-up of cellulase production and utilization. In: Aubert JP, Beguin P, Millet J
     (eds) Biochemistry and Genetics of Cellulose Degradation. FEMS Symp 43. Academic
     Press London p 71
198. Biely P (1985) Trends Biotechnol 11: 286

199. Teeri TT, Jones TA, Kraulis P, Rouvinen J, Penttilä M, Harkki A, Nevalainen H, Vanhanen S, Saloheimo M, Knowles JKC (1990) Engineering *Trichoderma* and its cellulases. In: Kubicek CP, Eveleigh DE, Esterbauer H, Steiner W, Kubicek-Pranz EM (eds) *Trichoderma* Cellulases: Biochemistry, Physiology, Genetics and Applications. Royal Chemical Society Press Cambridge p 156
200. Claeyssens M, Tomme P (1990) Structure-function relationships of cellulolytic proteins from *Trichoderma reesei*. In: Kubicek CP, Eveleigh DE, Esterbauer H, Steiner W, Kubicek-Pranz EM (eds.) *Trichoderma* Cellulases: Biochemistry, Physiology, Genetics and Applications. Royal Chemical Society Press Cambridge p 1

Note added in proof:

While this review was in press, the gene for *T. reesei* β-glucosidase was cloned and sequenced (Barnett CC, Berka RM and Fowler T (1991) Bio/Technology 9: 562)

Bioconversion of Cellulosic Materials to Ethanol by Filamentous Fungi

Ajay Singh, P. K. R. Kumar and K. Schügerl
Institut für Technische Chemie, Universität Hannover, Callinstraße 3,
3000 Hannover 1, FRG

The microbial production of ethanol and other solvents from renewable biomass is an attractive alternative to fuels and basic chemical feedstocks. Considerable efforts have been made in the past 10 years to improve the production of altenative fuel chemicals by various biological systems. Much current interest is focussed on the processes based on cellulosic and hemicellulosic feedstocks, the hydrolyzates of which contain a complex mixture of sugars. The technology to convert hexoses to ethanol is well established, however, conversion of pentoses and other sugars poses problems. Filamentous fungi belonging to the genera *Fusarium, Monilia* and *Neurospora* have been identified as potential organisms in recent years, that can convert cellulose directly to ethanol. Some species belonging to *Fusarium, Mucor* and *Paecilomyces* were also found to efficiently convert xylose to ethanol with high yields. Some fungal strains exhibited relatively higher ethanol and sugar tolerance. However, the major disadvantage with mycelial fungi for ethanol production is the slow bioconversion rate when compared to yeast. Potential bioethanol producing fungal strains, production of extracellular polysaccharases (cellulases and xylanases) and bioconversion of various carbohydrates are reviewed. The factors playing a significant role in determining culture variables and performance in lignocellulosic hydrolysate are discussed.

Advances in Biochemical Engineering/
Biotechnology, Vol. 45
Managing Editor: A. Fiechter
© Springer-Verlag Berlin Heidelberg 1992

1 Introduction

The interest in transforming vast reserves of renewable plant biomass to fuels and chemicals was generated because of the cost and availability of fossil fuel. Bioconversion of renewable cellulosic materials to various chemicals is an attractive approach in a world facing rapid depletion of fossil reserves. Moreover utilization of these materials in producing useful chemicals also helps to prevent overwhelming pollution problems. Any waste stream containing carbohydrates has the potential for generation of useful chemicals. Enzymatic hydrolysis of cellulose to glucose and its subsequent biological conversion to various solvents by microbial species is one of the processes under consideration.

Cellulose and hemicellulose are the two most abundant organic compounds in the biosphere. They are comprised of more than 60% plant biomass (Table 1) and occur in agricultural, forestry, fruits and vegetable waste processing. Cellulose has a high molecular weight; insoluble; β-1,4-linked polymer of glucose, whereas hemicelluloses are alkali-soluble heteropolymers of xylose, arbinose, galactose, mannose and glucuronic acid. It is necessary to utilize all these materials present in plant biomass to develop an economically feasible process.

Biomass can be converted to valuable chemicals either by thermochemical or biological means [1–10]. A thermochemical process, requires high temperature and pressure, and produces a complex mixture of products. On the other hand, a bioconversion process using microorganisms operates at a lower temperature and produces very few by products. Presently ethanol is produced either by catalytic hydration of ethylene or by employing yeast. Production of ethanol from biomass via microbial conversion is becoming increasingly attractive because of its several uses [11]: (1) potable ethanol in beer, wine, and in distilled beverages such as whisky, brandy rum etc.; (2) solvent ethanol for laboratory and pharmaceutical purposes; (3) as an antiseptic; (4) as a cosurfactant in oil-water microemulsions; and (5) as a fuel for automobiles, usually admixed with gasoline. Increase in the CO_2 level in the atmosphere and its possible effects has also prompted studies on the use of ethanol as automobile fuel. Photosynthesis ethanol production ethanol combustion cycle is advantageous since there is no net CO_2 generation [3]. Substitution of ethanol for gasoline would reduce air pollution [11].

Tabelle 1. Composition (%) of lignocellulosic substrates

Substrate	Hexosans	Pentosans	Lignin	Ref.
Bagasse	33	30	29	[67]
Birch	40	33	21	[33]
Corn stover	42	39	14	[67]
Groundnut shells	38	36	16	[67]
Oat straw	41	16	11	[68]
Pine	41	10	27	[33]
Rice straw	32	24	13	[33]
Wheat straw	30	24	18	[67]

The technology for bioethanol production from agricultural residues includes the substrate pre-treatment, substrate hydrolysis, biological conversion and product recovery. In the entire process, the most important properties for the choice of the microbial system are the range of substrates utilized and the ethanol yields. A wide variety of microbial species are now known to convert carbohydrates into many solvents such as ethanol, acetic acid, acetone, butanol etc. Organisms which have received more attention include several groups of yeast, a broad range of thermophilic and mesophilic bacteria and a few moulds [12–19]. The bacterium *Zymomonas mobilis* has been found to have the most potential for industrial ethanol production [20–26]. These organisms are capable of converting several hexose sugars including glucose, fructose, galactose, mannose, sucrose and maltose [27]. *Saccharomyces* spp. are the most intensively studied organism and have been used for centuries to produce alcoholic beverages. Several strains of yeasts have also been reported with the ability to convert starch, lactose and inulin [28–32]. However, the views of workers, on the ability of yeasts to produce ethanol from aldopentoses, were contradictory until recently [33, 34]. After a search for such yeasts, several appear to be of potential belonging to genera *Pachysolen*, *Candida* and *Pichia* [35–44].

Since the estimates of conventional feedstock cost for production of ethanol are in the conversion range from ca. 30–70% of the ethanol selling price, much current interest has been focussed on improving processes based on cellulosic and hemicellulosic feedstocks [45]. The simultaneous saccharification and biological conversion of cellulose, cellulase enzyme, yeast, and nutrients in the same reactor has been attempted in several laboratories to improve the bioconversion process of renewable cellulosic resources [46–53]. Continuous production of ethanol from cellulose was efficiently achieved by co-immobilizing yeast and cellulase in glass fibre [54]. However, separate steps for cellulase production and concentration are required. An alternative approach has been the direct conversion of cellulosic substrates in which the single organism carries out both hydrolysis and the biological conversion into useful chemicals in a single reactor. This direct approach eliminates the need for separate enzyme production and hydrolysis reactors. Thus during enzyme production, cellulose can be directly converted to ethanol by bacterial [55–60] and fungal [61–66] strains. Cellulose conversion by bacteria and fungi produces a yield comparable to that of yeasts, however, overall productivities are lower. But the advantage of using fungi or bacteria is that they produce cellulolytic and hemicellulolytic enzymes in addition to ethanol, whereas yeast does not metabolize cellulose.

A considerable progress has been made in obtaining ethanol from various carbohydrates using yeast and bacteria. Excellent reviews are available that deal with yeast and bacterial bioconversion processes [2, 6, 28, 34, 45, 69–74]. The present review will concentrate on the available literature concerning the direct bioconversion of cellulose, as well as sugars present in the hydrolyzate by filamentous fungi. It covers the description of the various potential fungal strains, and the aspects of some recent research progress on the bioconversion of cellulose and other carbohydrates, and also the evaluation of factors that affect the ethanol production by these fungal strains. The distinguishing features of filamentous fungi for ethanol production in comparison to yeast and bacterial systems are identified.

2 Potential Fungal Strains for Ethanol Production

Several fungal strains are known to possess enzyme systems for ethanol production [75], but no attempt has been made for developing a process technology. Several filamentous fungi belonging to the genera *Fusarium* [64–66, 76–82], *Rhizopus* [83, 84], *Mucor* [13, 76], *Neurospora* [13, 62, 63], *Monilia* [61], *Polyporus* [76, 85] and *Paecilomyces* [86] are known to convert glucose as well as xylose to ethanol (Table 2). It has probably been suspected by a number of investigators who noted a rather distinctive odour of ethanol while examining the growth of certain fungi [76]. Ethanol was observed to be a rather common metabolite when growth of a number of phycomycetes in submerged cultures was evaluated [76].

The ability of *Fusarium* sp. to produce ethanol from sugars was reported about 70 years ago [87, 88], however, detailed studies have only recently been undertaken [33, 78, 80, 89]. The specific growth rate of *F. oxysporum f. lini* (Bolley) declined with increasing cell mass. However, this organism was able to convert xylose efficiently with a yield of 0.41 g per g xylose utilized [80]. Suihko and Enari [77] tested 26 strains for their ability to produce ethanol from glucose and xylose. Except for the strain *F. tricinctum* VTT-D-72014, all the *Fusarium* strains were found to utilize both sugars. *F. oxysporum* VTT-D-80134 was the best strain with

Table 2. Ethanol producing fungal species

Organism	Features	Ref.
Aspergillus niger	Conversion of glucose to ethanol and acetic acid	[76]
Fusarium lini	Conversion of glucose to ethanol and acetic acid	[76]
Fusarium spp.	Efficient conversion of xylose and glucose to ethanol	[77, 81]
F. oxysporum F3	Direct conversion of cellulose into ethanol	[64, 65]
F. oxysporum DSM 841	Efficient direct conversion of waste cellulose to acetic acid	[66]
F. oxysporum (*lini*)	Efficient conversion of xylose	[80]
Monlia sp.	Direct conversion of cellulose to ethanol	[61]
Mucor sp.	Conversion of glucose to ethanol, acetic acid and oxalic acid	[76]
Neurospora crassa	Direct conversion of cellulose/ hemicellulose into ethanol	[62, 63]
Paecilomyces sp.	Efficient ethanol production from all the sugars present in the hydrolysate of plant biomass	[86]
Polyporus anceps	Conversion of starch, sucrose and fructose into ethanol/acetic acid	[76, 85]
Rhizopus oryzae	Efficient conversion of glucose to ethanol and lactic acid	[84]

an ability to yield 0.46 g ethanol per g glucose and 0.42 g ethanol per g xylose present initially [77]. Perlman [76] tested 6 *Mucor* strains for their ability to convert glucose. Each of these strain was found to be able to produce at least $1 \, g \, l^{-1}$ ethanol. *Mucor* 101 and *Mucor* 105 were able to produce 7 to $66 \, g \, l^{-1}$ ethanol from a range of sugars [13]. *Paecilomyces* sp. NF1 was found to have potential characteristics regarding ethanol bioconversion [86]. It was capable of converting all the major sugars present hydrolysate of plant biomass. So far the highest reported yield ($76 \, g \, l^{-1}$) from xylose is with this organism.

A few fungal strains have been reported as having the ability to convert cellulose directly to ethanol. *Monilia* [61], *Neurospora crassa* [62, 63], and *Fusarium oxysporum* [64–66] were found to have the potential in carrying out both hydrolysis and conversion of cellulose into ethanol/acetic acid. Ethanol production is usually accomplished by placing aerobically grown mycelia under anaerobic conditions. However, the process of ethanol production by filamentous fungi is hampered by the need for a long time for high conversion [33, 34]. This is quite understandable in view of developing a process technology where yeasts are readily adaptable to rapidly utilize sugars and convert them into ethanol. However a fungal system might be of interest due to its ability of growing on cellulose and natural plant biomass which a yeast system usually cannot.

3 Glucose and Xylose Metabolism in Fungi

A number of metabolic pathways are well documented in various microorganisms for the synthesis of ethanol from glucose [11]. Glucose is converted to ethanol in fungi and yeasts by the Embden-Meyerhof-Parnas (EMP) pathway which occurs in the cytosol [27]. Theoretically, 0.51 g ethanol and 0.49 g CO_2 are yielded from 1 g glucose. However, the real ethanol and CO_2 yields are 0.46 g and 0.44 g from 1 g glucose, since 0.10 g glucose is metabolized for biomass production. Both facilitated and active uptake mechanisms are responsible for the transport of sugars into the cell [28], however, metabolic control of carbohydrate catabolism is very complex and still not clearly understandable [90]. The two major controls of glucose catabolism discovered in yeasts are; (1) 'Pasteur effect', the inhibition of glycolysis by oxygen, and (2) 'Crabtree effect', repression of respiration and deregulation of the glycolytic pathway [28].

The first step of xylose catabolism is its conversion to xylulose. In bacteria, it takes place by the direct isomerization catalysed by xylose isomerase. In *Penicillium chrysogenum*, a sequence of enzymes in the initial steps of pentose metabolism was observed that differs from xylose isomerization in bacteria [91, 92]. These enzymes were common in yeast and filamentous fungi. In this oxido-reductive pathway, xylose is first reduced to the xylitol in the presence of NAD(P)-linked xylose reductase, which is then reoxidized by NAD(P)-linked dehydrogenase to give xylulose (Fig. 1). It has been assumed that this oxido-reductive pathway is common among fungi [93]. Both the enzymes involved, xylose reductase and xylitol dehydrogenase, were found to be inducible and relatively specific for the D-xylose and xylitol in *F. oxysporum*, whereas D-xylose isomerase was not detected.

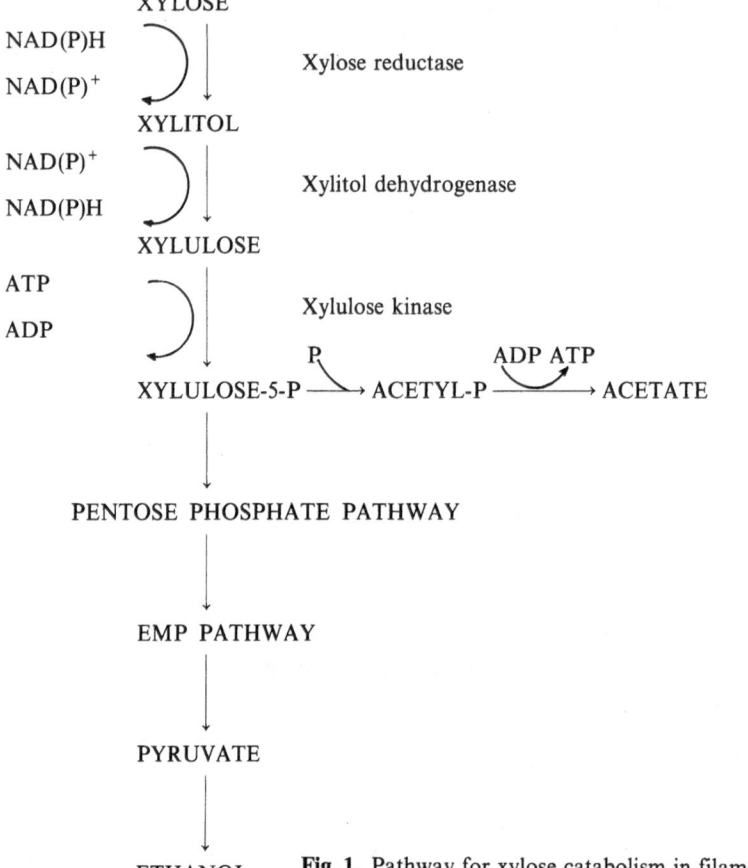

Fig. 1. Pathway for xylose catabolism in filamentous fungi

However, the enzymes of both pathways have been demonstrated in *Candida utilis* where xylose isomerase was found inducible and isomerization pathway occurred when xylose as the carbon source was alone in the medium [94, 95]. When both pentoses and hexoses were supplied to the medium, the conversion was carried out through oxido-reductive pathway. Indirect evidence for the presence of xylose isomerase in *Rhodotorula gracilis* and other yeast like organisms has also been reported [96, 97].

A further step in the D-xylose catabolism is believed to be phosphorylation of D-xylulose to D-xylulose-5-phosphate [2, 98, 99]. The steps after formation of xylulose phosphate appear to use a combination of pentose phosphate and EMP pathways to the key intermediate pyruvate. This process is followed by the conversion of pyruvate to ethanol by the action of pyruvate decarboxylase and alcohol dehydrogenase. D-xylulose-5-phosphate, after converting into D-glyceraldehyde-3-phosphate via the enzymes of the pentose phosphate cycle, is also thought to be incorporated into biomass and oxidized to CO_2 and water via respiratory processes [34]. Finally, 3 moles pent(ul)ose phosphate are converted into 5 moles pyruvic acid with a net synthesis of 5 moles ATP and 5 moles reduced pyridine nucleotide [2]. The metabolic pathways in pentose conversion are more complicated than the hexose conversion. This may be the possible explanation for the slow conversion rate of pentoses as compared with hexoses [82].

Resting cells of *F. oxysporum* produce a significant quantity of acetate, whereas growing cells produce higher amounts of ethanol and less acetic acid [33]. According to the pathway suggested, equimolar amounts of ethanol and acetate should be synthesized [82, 100]. The difference observed between the ethanol yields in resting and growing cells suggest that either growing cells follow a different metabolic pathway for the conversion or that they are able to reduce acetate to ethanol [2]. In growing fungal cultures, ethanol was found to be metabolized to acetic acid [76]. The detection of oxalic acid in this bioconversion suggested that acetic acid was further metabolized to oxalic acid. Under certain conditions, growing cultures of *Fusarium* was found to utilize acetate as the sole carbon source for growth [101]. The ability of growing cells of Fusaria to metabolize ethanol also suggest the preferential dissimilation of acetic acid over the ethanol since the latter is produced in the same yield in both bioconversions. This preferential dissimilation might be due to the non-availability of ethanol for growth by proliferating cells so long as sugar was present.

4 Characteristics of Ethanol Producing Fungi

4.1 Simultaneous Saccharification and Biological Conversion

The bioconversion of cellulose into ethanol with conventional methods is usually achieved in two steps; first being the enzymatic saccharification of the polysaccharide to monosaccharide; and secondly the bioconversion of monosaccharides into ethanol. A combination of enzymatic hydrolysis and ethanol production in the same reactor has been attempted using different cellulases and ethanol producing microbial species to improve process efficiency [46–53]. The production of ethanol from cellulose in a simultaneous saccharification and biological conversion process alleviates the problem of end product inhibition, since glucose does not accumulate in this system and is converted to ethanol immediately following saccharification [46].

The enzymatic activity and ethanol formation abilities of *Fusarium oxysporum* makes it ideal for a simultaneous saccharification and biological conversion process [82]. Birch wood containing high xylose content, was used for the simultaneous saccharification and biological conversion test [82]. The birch chips were steam-treated at 185 °C for 30 min. Cellulase from *Trichoderma reesei* and either *Saccharomyces cerevisiae* or *Fusarium oxysporum* were used for the simultaneous saccharification and biological conversion process. Rates of hydrolysis and bioconversion were studied by varying the amount of *T. reesei* cellulase, since the polysaccharases produced by *F. oxysporum* were not sufficient to hydrolyze the substrate alone. The growth and ethanol production rate of *Fusarium* were slower than that of *Saccharomyces*, however, the final yield was higher because of the ability of *Fusarium* to convert pentoses. After a lag phase during the first day, all the glucose as well as most of the other reducing compounds were consumed by *F. oxysporum*. On the other hand, pentoses accumulated in the medium with the yeast bioprocess. The viscous nature of the media also affected the regulation of the aeration rate [82].

Paecilomyces sp. NF1 possess several favourable characteristics for the simultaneous saccharification and biological conversion process [86]. This fungus was able to produce about $60 \, g \, l^{-1}$ ethanol in simultaneous saccharification and biological conversion of either $150 \, g \, l^{-1}$ cellulose or $100 \, g \, l^{-1}$ cellulose with $50 \, g \, l^{-1}$ xylan using *T. reesei* cellulase and *A. niger* hemicellulase. *Paecilomyces* was also found to utilize all the sugars present in an acid hydrolyzate of wheat straw.

4.2 Direct Bioconversion of Cellulosic Materials

The direct bioconversion of cellulosic materials by fungal cultures is an attractive approach in which a single organism carries out the hydrolysis of substrate under aerobic conditions and produces ethanol under semianaerobic conditions. Production of extracellular hydrolytic enzymes and ethanol by some potential fungal strains are discussed in this section.

4.2.1 Production of Polysaccharases

Cellulose is degraded by a multienzyme complex involving at least three enzymes: endo-1,4-β-D-glucanase (EC 3.2.1.4), exo-1,4-β-D-glucanase (EC 3.2.1.91), and a β-D-glucosidase (EC 3.2.1.21) also called as cellobiase [102, 103]. Endoglucanase cleave β-1,4 chains of cellulose randomly, whereas exoglucanase releases cellobiose or glucose units from the non-reducing end of the cellulose polymer. For the complete hydrolysis of insoluble cellulose, synergistic action among these components is required. Cellulases are produced by bacteria, actinomycetes, fungi [104], algae, myxobacteria [105], basidiomycetes and also by some higher forms like molluscs and insects [104–107].

Several microorganisms have been reported to synthesize cellulolytic enzymes when grown in a medium containing cellulose or a cellulase inducer [108], however, they cannot convert sugars to ethanol. On the other hand, many organisms are able to convert saccharoses to ethanol but they lack the genetic information to produce the polysaccharases for the hydrolysis of cellulose [61]. A few filamentous fungal strains have been reported recently that are able to hydrolyze and convert cellulose directly to ethanol. Table 3 shows a comparison of maximum enzyme activities produced by these strains.

Monilia sp, isolated from bagasse compost was found to utilize many polysaccharides including cellulose and produce ethanol, and cellulase and xylanase enzymes [61]. Cellulase activity was detectable in the culture medium after 48 h of growth. Interestingly, when cellulose was used as a substrate, both cellulase and

Table 3. Production of extracellular hydrolytic enzymes by ethanol producing fungi

Organism	Activity (U ml^{-1})				Ref.
	Exoglu-canase	Endoglu-canase	β-Glu-cosidase	Xylanase	
Fusarium oxysporum F3	0.38	33.15	1.05	nd	[64]
Fusarium oxysporum 841	1.5[a]	3.2	4.9[b]	28.5	[66]
Monilia sp.	nd.	7.8	nd	5.0	[61]
Neurospora crassa	0.4[c]	6.4	0.5	nd	[63]

[a] activity against Avicel; [b] activity against cellobiose; [c] activity against filter paper; nd not determined

xylanase activities were produced by *Monilia* sp. Similar results were obtained in our laboratory with *Monilia brunnae* [66]. However, only xylanase was detected when xylan was used as the sole carbon source [61]. *Streptomyces* and *Chaetomium* species have also been reported to produce xylanase in the presence of xylan as the carbon source and inducer [109, 110]. On the other hand, xylan is not necessary for a few organisms to induce xylanase, as high xylanase activities were produced with pure cellulose or glucose [111].

A number of *Neurospora* cultures were screened for cellulase and ethanol production [62, 63]. Detailed studies on the direct bioconversion with two promising strains of *N. crassa* NCIM 870 and 1021 were made by these workers. Studies on enzyme production were carried out in aerobic precultures using alkali-treated cellulose powder, untreated cellulose powder and wheat bran in two different media; yeast extract-malt extract-peptone, and Mandels and Weber medium [106]. Cellulose activities were detected as early as after 48 h in both media. Maximum activities against filter paper and Carboxymethyl cellulose (CMC) were obtained after 7 days of incubation, whereas β-glucosidase activity was maximum after 10 days [62]. The optimum pH for the production of cellulase by *N. crassa* was 7.0 and under assay condition the optimum pH was 5.0. The optimal temperature for the enzyme assay was 50 °C, whereas enzyme production was carried out in a shaken flask at 28 °C. The production and purification of xylanase enzyme by *N. crassa* 870 was also evaluated [112]. The extracellular enzyme activities were very poor with glucose as the substrate, however, a significant amount of xylanase activity was synthesized using a cellulose powder carbon source. Xylanase activity reached a maximum (14 U ml^{-1}) on the 4th day using commercial xylanase substrate. On the other hand, cellulase activities were higher when cellulose powder was used in combination with wheat or wood. The optimum pH was 5.0 and xylanase production decreased above or below this pH value. The purified endoxylanase of *N. crassa* was able to hydrolyze xylan (Fluka) by 66% in 4 h. The xylosidase activity, however, was very low which might have affected the incomplete hydrolysis of xylan [112].

Fusarium species are regarded as potential decomposers of plant biomass e.g. cellulose, hemicellulose, pectins and lignins [81, 89]. Christakopoulos et al. [64] screened and selected *Fusarium oxysporum* strains on the basis of their ability to excrete high level of cellulase and β-glucosidase using straw as the carbon source. Among the three strains selected, *F. oxysporum* F3 produced higher amounts of cellobiohydrolase, activity against CMC and β-glucosidase than F1 and F2. Maximum activities in culture media were found on the 7th day of incubation. Of the three main enzymes involved in cellulose saccharification, β-glucosidase was found to be the key enzyme in direct conversion of cellulose to ethanol by *F. oxysporum* [65]. Ethanol production was not affected when the activities of cellobiohydrolase and CMCase were varied at the beginning of non-aerated growth. However, ethanol production was affected significantly by adjusting β-glucosidase at the optimum level of 0.7 to 0.8 U ml^{-1} in the growth medium [65]. The maintenance of β-glucosidase activity at the optimum level, during the beginning of the bioprocess, also decreased significantly the cellulose bioconversion time to ethanol. An important point noticed (unpublished results of the authors) and by Christakopoulos et al. [64, 65], was that the pH of the aerated culture (enzyme production phase) had a marked effect on ethanol yield. Adjustment of the initial pH to 4.5 in non-aerated growth established an optimum pH for the functioning of *F. oxysporum* F3 and ethanol production, whereas pH 4.5–5.0 of non-aerated cultures of *F. oxysporum* 841 (unpublished results) resulted in enhanced productivities.

Four strains of *F. oxysporum* were screened for their ability for the simultaneous production of hydrolytic enzymes and ethanol/acetic acid from cellulosic substrates [66]. All the strains were found to utilize cellulose and produce extracellular cellulase and xylanase enzymes. *F. oxysporum* 841 was found to be a potential mould for direct conversion of cellulose to ethanol/acetic acid. Though other species, e.g. *F. oxysporum* 2018, 62287 and 62291 also produced comparable amounts of cellulase and xylanase enzymes, only traces of ethanol and acetic acid were detected in non-aerated cultures. Maximum cellulase activity was observed after 96 h of bioconversion process. *F. oxysporum* VTT-D-80134, however, was not found to produce sufficient cellulolytic and xylanolytic activities to convert cellulose or hemicellulose directly into ethanol [81].

Uncompetitive inhibition of cellulase enzyme by ethanol has been observed in the case of simultaneous saccharification and biological conversion of cellulose [51]. However cellulases of ethanol producing fungi were found to be stable in the presence of ethanol, where levels of enzyme were maintained throughout the semianaerobic phase of bioconversion process [65, 66].

4.2.2 Production of Ethanol

Several biological systems have been identified in the recent past which can convert cellulose and hemicellulose, present in natural crop residues, directly to ethanol. Thermophilic bacterial species like *Clostridium thermocellum* and *C. thermohydrosulfuricum* in mono- and co-cultures were found to have a potential for direct the conversion of cellulose into a variety of products, such as ethanol, acetic acid and lactic acid [55–58]. The potential of producing ethanol from lignocellulosic materials by thermophilic bacteria have been recently reviewed by Lynd [113]. *N. crassa*, *Monilia* and *F. oxysporum* are potential mould cultures for the direct bioconversion of cellulose.

More than 70% of the Solka-floc and 60% of Avicel were converted to ethanol in a single step by a *Monilia* sp. [61]. Ethanol yield of about 16 g l^{-1} and 12 g l^{-1} were obtained from Solca-flok and Avicel, respectively, in 8 d under anaerobic conditions. The results also indicate that *Monilia* sp. was also able to utilize hemicellulose and pectic materials in addition to cellulose. This is an important feature because hemicellulose is a major component of agricultural residues.

The direct bioconversion of alkali-treated cellulose powder (ATCP) and stream-treated bagasse to ethanol by two strains of *N. crassa* was investigated [62]. A conversion of 80–90% of ATCP with a yield of 18 g l^{-1} ethanol and 60% conversion of cellulose plus hemicellulose in bagasse by *N. crassa* was achieved in 10 days. The conversion of D-xylan was 58% in 6 days, whereas alkali-treated bagasse was converted to ethanol in 4 days by this organism with a conversion of 90% (based on 60% convertable carbohydrates in bagasse). Increase in bagasse substrate concentration (5%) resulted in increased ethanol yield (11 g l^{-1}) corresponding to 75% conversion in 6 days.

A high cellulase producing strain of *F. oxysporum* F3 was tested on a number of carbon sources for its ability to convert them directly to ethanol [64]. Among the various carbon sources tested, cellulose was the best for ethanol production by this strain. Maximum ethanol concentration (9.6 g l^{-1}) with a theoretical yield of 89.2% was achieved in 6 days using 20 g l^{-1} cellulose. The production of ethanol was increased with a substrate concentration from 20 to 50 g l^{-1}. A considerable reduction in the bioconversion time was observed when β-glucosidase activity was

Table 4. Direct conversion of cellulosic material by fungi

Organism	Substrate (g l^{-1})		Time (d)	Ethanol (g l^{-1})	Acetate (g l^{-1})	Ref.
Fusarium	Cellulose	(20)	6	9.6	—	[64]
oxysporum	Cellulose	(50)	6	14.5	—	
F. oxysporum	Avicel	(10)	7	1.5	2.5	[66]
DSM 841	Potato pulp	(30)	6	0.1	4.7	
	Avicel	(20)	6	3.6	4.8	[127]
	Potato pulp	(50)	6	3.6	12.0	
Monilia sp.	Solka floc	(50)	8	16	—	[61]
	Avicel	(50)	8	12	—	
Neurospora	ATCP	(20)	7	9.9	—	[63]
crassa	ATCP	(50)	7	9.9	—	
	Solka floc	(20)	9	6.9	—	
	Avicel	(20)	7	9.9	—	
	Avicel	(50)	7	9.9	—	
	Bagasse	(50)	6	11	—	

ATCP = alkali treated cellulose powder

adjusted to the optimum level at the beginning of anaerobic growth [65]. Thus maximum ethanol yields were obtained in 3–4 days, approximately half the normal bioconversion time [61, 63, 65].

Equimolar concentrations of ethanol and acetic acid were obtained when *F. oxysporum* 841 was tested for its ability of direct conversion in shaken flasks [66]. This strain had the potential for the production of acetic acid from cellulosic materials. Acetic acid yield of $2.5 \, g \, l^{-1}$ from Avicel and $4.7 \, g \, l^{-1}$ from potato waste (a major byproduct from starch industries) was obtained in a single step process indicating the potential of this strain for converting cellulosic substrates into acetic acid.

4.3 Whole Cell Immobilization

The immobilization of whole cells provides a means for the entrapment of multistep and cooperative enzyme system present in the intact cell, repetitive use and improved stability. This technique is also advantageous in the separation of bioproducts from cell mass in a continuous bioconversion process [114, 115]. The other advantages of immobilized growing cells include: (1) protection of cells against unfavourable environmental factors; (2) changes in the permeability of the cells; (3) reduced inhibition by substrate and product; (4) reusability; and (5) faster removal of end product.

To date, there is only one report on the use of immobilized fungal cell for the production of ethanol [113]. Two mould cultures, *Mucor* sp. and *F. lini* were compared with *S. cerevisiae* for their ethanol producing ability. Immobilization was attempted as a means to increase the volumetric rate of ethanol production. Spores of *Mucor* sp. were entrapped in alginate gel. The size of the alginate pellets was 2.5 mm diameter with a cell density of $0.33 \, g \, ml^{-1}$. A spore population of 10^5 was measured by direct microscopic count. *F. lini* was grown upon the ground corn cobs (3.2 mm diameter). In shaken flasks, both fungal cultures produced ethanol from 10% xylose at lower rates compared to *S. cerevisiae* which had the best ethanol production rate, $0.3 \, g \, l^{-1} \, h^{-1}$.

4.4 Continuous Culture

The continuous culture technique has been used successfully in bioconversions involving filamentous moulds [80]. This system may allow relatively higher ethanol production while enjoying some of the advantages of a continuous system. Different residence times for liquid and insoluble substrate in a continuous system are relatively easy to achieve, moreover cells attached to the substrate are also retained in this system [113].

Continuous cultivation of *F. oxysporum* (*lini*) was carried out on the basis of physiological and yield data obtained in batch cultures [80]. It was operated at a dilution rate of $0.0112 \, h^{-1}$ under steady state conditions with a feed concentration of 1% (w/v) xylose. At this dilution rate, 98.5% substrate utilization was achieved compared to 95% substrate utilization at a dilution rate of $0.018\,h^{-1}$. A cell recycle system was employed to increase the concentration of mould mycelium in the bioreactor and to obtain higher conversion rates. Continuous cultivation was operated with an increase in the dilution rate by a factor of 2.6, i.e. at $0.029\,h^{-1}$. With this system, 98.4% utilization of the 1% xylose feed was achieved. The ethanol productivity enhanced from $0.04\,g\,l^{-1}\,h^{-1}$ to $0.07\,g\,l^{-1}\,h^{-1}$, a factor of 1.75. Cell recycle has been reported to increase both the bioconversion rate and ethanol yield, which is probably due to the increase in cell density as well as increasing the limitation of the culture growth [44].

A continuous culture system using alginate immobilized whole cells of *F. lini* and *Mucor* sp. [114] was carried out in a 500 ml column reactor (6 cm × 18 cm) at a dilution rate of 1 day^{-1}. Ethanol at low concentration was produced by both organisms. Accumulation of CO_2 followed by possible reduced efficiency of the column reactor was noted by these workers. However, an increase in the conversion rate by *Mucor* sp. was achieved when pre-isomerized xylose solution was used as substrate.

The results obtained on ethanol production, in a continuous culture of immobilized mould system were found unsatisfactory, because of the slow conversion rate with pentose as substrate, thus improvements are still needed.

5 Factors Affecting Ethanol Production by Fungi

5.1 Cultivation Conditions

It was observed in most of the cases that the fungal system would not grow under anaerobic conditions, and would not produce ethanol under aerobic conditions. Thus, the fungus was grown under aerobic conditions and then the mycelia were placed in semianaerobic conditions for ethanol production. Wu et al. [86] used sealed Bellco tubes for studying bioethanol production by *Paecilomyces* sp. Experiments with various species of *Fusarium* were conducted in a 250 ml Erlenmeyer flask, equipped with a water trap and a tube with an injection needle through the rubber cap [77]. *Monilia* sp. was grown aerobically for 24 h, thereafter, additional substrates were added and flasks were incubated under non-aerated condition for the conversion of cellulose to ethanol. Rao et al. [62] suggested two stages for the ethanol production by *N. crassa*: (a) aerobic phase, in which the mould is grown aerobically for 48 h; and (b) non-aerated phase, in which the contents of growth flasks are transferred to a special flask, with a capillary opening at the top to minimize the diffusion of oxygen and to allow CO_2 to escape. A similar technique was adapted with *F. oxysporum* by Christakopoulos et al. [64]. Perlman [85], who carried out bioconversion studies with *P. anceps* in an Erlenmeyer flask equipped with twohole rubber stoppers. Sterile water saturated CO_2 free air was passed through one of the openings, and left the flask through the other. The exhaust tube was connected to a flask of 2 N KOH and air flow was controlled with a small valve and a manometer.

5.2 Carbon Sources

A wide variety of sugars were used by different workers for the production of ethanol by fungi (Tables 5–7). *Rhizopus oryzae* was able to produce 6.2 g l^{-1} at a glucose consumption rate of 7.2 g l^{-1} h^{-1} during a 35 h period [84]. This fungus was found to be able to produce simultaneously 9.66% lactic acid, the yield being 70–75% based on the glucose consumed. Ethanol, acetic acid, oxalic acid and CO_2 were apparently the major products of carbohydrate dissimilation by *Polyporus anceps* [85]. Approximately one mmol ethanol was formed per mmol glucose utilized in the process [85].

Table 5. Bioconversion of glucose by various fungal species

Organism	Substrate (g l^{-1})	Time (d)	Ethanol (g l^{-1})	Acetate (g l^{-1})	Ref.
Aspergillus niger	10	3	1.1	1.2	[76]
Fusarium lini	10	6	2.3	0.7	[76]
F. avenaceum	50	4	13–20	—	[77]
F. clamydosporium	50	4	21	—	[77]
F. culmorum	50	4	14	—	[77]
F. elegans	50	4	14	—	[77]
F. graminearum	50	4	12–21	—	[77]
F. poae	50	4	18	—	[77]
F. solani	50	4	6–23	—	[77]
F. sporotrichiella	50	4	13	—	[77]
F. sambucium	50	4	20	—	[77]
F. sporotrichoides	50	4	22	—	[77]
F. tricinctum	50	4	13	—	[77]
F. oxysporum	50	4	22–23	—	[77]
F. lycopersici	160	4	28	—	[13]
Fusarium F5	160	4	38	—	[13]
F. oxysporum	20	6	8.2	—	[64]
F. oxysporum	20	4	4.4	4.6	[127]
Monilia sp.	50	7	22	—	[61]
Mucor heimalis	10	3	2.6	0.3	[76]
M. mucedo	10	6	1.9	1.1	[76]
M. piriformis	10	6	2.3	0.6	[76]
M. pusillus	10	3	1.1	0.5	[76]
M. romannianus	10	3	2.3	0.3	[76]
M. varians	10	3	2.1	0.4	[76]
Mucor 101	160	4	66	—	[13]
Mucor 105	160	4	65	—	[13]
Neurospora crassa	10	3	1.6	0.3	[76]
	20	4	22	—	[63]
Paecilomyces sp.	100	7	40.9	—	[86]
Polyporus anceps	20	10	1.7	3.1	[55]
P. anceps	10	6	3.8	0.2	[76]
Rhizopus oryzae	133	1.5	6.2	—	[84]

Table 6. Bioconversion of xylose by various fungal species

Organism	Substrate $(g\,l^{-1})$	Time (d)	Ethanol $(g\,l^{-1})$	Ref.
Fusarium poae	50	4	13	[77]
F. avenaceum	50	4	8–12	[77]
F. clamydosporium	50	4	11	[77]
F. culmorum	50	4	9–12	[77]
F. elegans	50	4	4	[77]
F. graminareum	50	4	6–11	[77]
F. sambucium	50	4	13	[77]
F. solani	50	4	3–11	[77]
F. sporotrichiella	50	4	4	[77]
F. sporotrichoides	50	4	4	[77]
F. trincinctum	50	4	7	[77]
F. oxysporum	50	4	13–21	[77]
F. oxysporum (*lini*)	10	4	3.5	[80]
F. oxysporum	20	6	5	[64]
F. oxysporum	20	4	4	[127]
F. lycopersici	50	4	16	[13]
Fusarium F5	50	4	14	[13]
Monilla sp.	50	6	10	[61]
Mucor 101	50	4	8	[13]
Mucor 105	50	4	8	[13]
Neurospora crassa	20	6	6.7	[63]
Paecilomyces sp.	100	7	39.8	[86]
	200	7	73.5	[86]

The ability of *Mucor* and *Fusarium* strains for the production of ethanol was tested on various sugars viz. glucose, sucrose, xylose and sugar alcohol, xylitol [13]. *Mucor* 105 was not able to utilize sucrose, whereas the other three strains converted all the sugars to ethanol efficiently. However, the rates of D-xylose and xylitol assimilation and subsequently ethanol production were low when compared to D-glucose. D-Glucose was always the preferred substrate being utilized within the first day. When L-xylose, D-arabinose or L-arabitol were used as carbon sources no ethanol was produced. *F. oxysporum* (*lini*), in a pH controlled batch conversion of 1% xylose, was able to achieve 41% yield of ethanol [80]. An acid extract of wheat straw containing $2\,g\,l^{-1}$ glucose and $9\,g\,l^{-1}$ D-xylose was converted by this organism. Ethanol at a level of $3.2\,g\,l^{-1}$ was produced, giving a conversion of 29% assuming all the sugars were converted to ethanol.

The cellulolytic strain of *Monilia* sp. was able to produce ethanol from both D-glucose and D-xylose [61]. More than 40% D-xylose was converted to ethanol in 7 days. However, the conversion rate was slower than that of D-glucose. Similar results were obtained by Kumar et al. [66] with *F. oxysporum*. However, the ability of cellulolytic strains of fungal cultures to produce ethanol is of special significance, since xylose is a major constituent of cellulosic materials which is usually left unutilized, due to the lack of a suitable organism to convert it efficiently [61, 116]. Another cellulolytic strain of *F. oxysporum* has been found to convert successfully

Table 7. Bioconversion of other carbohydrates by various fungal species

Organism	Substrate (g l^{-1})		Time (d)	Ethanol (g l^{-1})	Acetate (g l^{-1})	Ref.
Fusarium	Sucrose	(100)	4	36	–	[13]
lycopersici	Xylitol	(50)	4	5	–	
Fusarium F5	Sucrose	(100)	4	44	–	[13]
	Xylitol	(50)	4	2	–	
F. oxysporum	Mannose	(50)	4	22	–	[81]
	Galactose	(50)	6	21	–	
	Fructose	(50)	7	21	–	
	Cellobiose	(50)	7	17	–	
	Sucrose	(50)	4	23	–	
	Ribose	(50)	4	3	–	
	Xylitol	(50)	6	8	–	
F. oxysporum	Cellobiose	(20)	6	8.9	–	[64]
	Cellobiose	(50)	6	17.5	–	
Mucor 101	Sucrose	(100)	4	11	–	[13]
	Xylitol	(50)	4	9	–	
Mucor 105	Xylitol	(50)	4	7	–	[13]
Paecilomyces	Fructose	(100)	7	39.9	–	[86]
sp.	Galactose	(100)	7	40.4	–	
	Mannose	(100)	7	38.1	–	
	Ribose	(50)	4	10.2	–	
	Arabinose	(50)	4	13.8	–	
	Cellobiose	(100)	7	40.2	–	
	Lactose	(50)	4	14.1	–	
	Maltose	(50)	4	21.1	–	
	Starch		7	39.8	–	
Polyporus	Sucrose	(20)	10	1.9	4.0	[85]
anceps	Fructose	(20)	10	1.9	2.3	
	Starch	(20)	10	0.8	2.0	
	Glycerol	(20)	10	1.0	0.9	

not only glucose and xylose but also cellobiose and cellulose to ethanol [64]. It produces ethanol in the range of 5.0 to 9.6 g l^{-1} with theoretical yield from 48.0 to 89.2%. Under anaerobic conditions, *F. oxysporum* was capable of producing ethanol with high yields from D-glucose, D-mannose, D-fructose and D-xylose, but a lower yield from xylitol and D-ribose was obtained [33]. Both D-glucose and D-xylose were converted to produce ethanol at a theoretical yield.

Neurospora crassa is another promising mould capable of producing ethanol from a wide range of carbon sources such as cellulose, xylose, arabinose, mannose, galactose and glucose [62, 63]. D-Mannose metabolized rapidly with 100% conversion in 2 days, while D-galactose gave 75% conversion in 6 days. The pentose sugars, D-xylose and D-arabinose, however, metabolized at a lower rate, yielding 65% and 44% conversion, respectively in 6 days. The biological conversion capacity of *N. crassa* on D-xylose was comparable to the yields obtained with

P. tannophilus [117, 118] and *Pichia stiptis* [119]. On the other hand, the bacterial process of xylose results in a mixture of different end products when ethanol concentration is low [2].

Paecilomyces sp. NF1 was capable of metabolizing a wide range of substrates including glucose, galactose, fructose, mannose, maltose, cellobiose, lactose, soluble starch, xylose, aranbinose and ribose [86]. The unique characteristic found in this fungus was its ability to produce ethanol at a level of 73 g l^{-1} from 200 g l^{-1} xylose. This is the highest reported yield of ethanol from a xylose bioconversion. Moreover, only traces of other chemicals were detected at the end of the bioprocess.

The ethanol yields reported are rather low from xylose bioconversion by yeast and bacteria, and the exact control of the oxygen level is required [28, 34]. Another difficulty encountered in xylose metabolism by yeast, is to achieve a complete conversion of xylose to xylulose economically [77]. Filamentous fungi convert xylose with a high ethanol yield, but the bioconversion rate is slow when compared to yeast and thus, requires further improvements.

5.3 Nitrogen Sources

The involvement of nitrogen in growth and metabolism is well known. Ammonium ions counteract the inhibition of phosphofructokinase by ATP thus stimulate glycolysis [120]. It also stimulates growth, decreases the intracellular level of NADH, derepresses glucose-6-phosphate dehydrogenase, and thereby enhances the activity of pentose phosphate pathway. A variety of inorganic [13, 76, 77, 84–86, 114] and organic nitrogen sources [61, 64] were evaluated by various workers (Table 8). Rao et al. [62] compared the ethanol production from cellulose by *N. crassa* in a medium containing inorganic (ammonium sulphate) and organic (peptone) nitrogen sources. Higher ethanol production was noticed in the medium containing organic nitrogen sources (peptone).

Table 8. Nitrogenous compounds used in biological conversion process by fungi

Nitrogenous compound	Concentration (g l^{-1})	Ref.
Ammonium nitrate	1.0	[33, 85]
Ammonium dihydrogen phosphate	2.0	[76]
Potassium nitrate	2.5	[86]
Peptone	5.0	[61–64, 66, 113]
Sodium nitrate	2.0	[13]
	3.0	[77]
	3.4	[113]
Urea	2.0	[84, 85]
Yeast extract	0.25	[76]
	1.0	[77]
	1.5	[86]
	3.0	[61, 62, 66, 113]
Potato protein liquor	40	[66]

Enari and Suihko [33] studied the effect of various nitrogen sources viz. $NaNO_3$, NH_4NO_3 and urea at different concentrations $(0-1.12 \text{ g N l}^{-1})$ on ethanol production by *F. oxysporum*. Increasing amount of nitrogen content in the culture medium increased the growth, ethanol production and xylose consumption till an optimum level. Nitrogen limitation led to the death of the cells. Equal conversion rates were observed with nitrogen concentration of 0.56 g l^{-1} for $NaNO_3$, 0.84 g l^{-1} for NH_4NO_3 and 1.12 g l^{-1} for urea. However, the most suitable utilization rates of nitrogen was obtained with $NaNO_3$ as the nitrogen source. Supplementation of aerated growth medium with wheat bran (0.2%) resulted in enhanced ethanol production and decreased the process time by half [65], however, reduction in conversion time could not be explained on the basis of the data available. The nitrogen sources used by various workers for the production of ethanol by fungi includes ammonium nitrate [76], urea [33, 84], ammonium dihydrogen phosphate [76], sodium nitrate [13, 77, 114], potassium nitrate [86], and peptone [61–66].

5.4 Minerals and Vitamins

In spite of the pronounced effect of minerals and trace elements with the biosynthesis of metabolites, negligible information is available with the fungal production of ethanol. Different combinations of mineral salts and trace elements, commonly used by various workers in ethanol production by fungi, are presented in Table 9. However, no systematic studies have been made on the effect of mineral salts on ethanol production by fungi. Trace elements such as Fe, Zn, Cu and Mn were found to increase the growth of *Polyporus anceps* [85], while Co was toxic at the concentration of $1-100 \text{ mg l}^{-1}$, and Mo exhibited no effect.

Little information is available regarding the effect of growth factors on ethanol production by fungi. Thiamin was found essential for the growth of *P. anceps* in submerged cultures [85]. As the concentration of thiamin in the medium increased, the quantity of mycelium formation increased as did the amount of glucose utilized and ethanol and acetic acid produced. To study the growth promoting effect on *P. anceps*, a number of other crude supplements including yeast extract, malt extract, liver extract, cornsteep liquor and malt sprout were added into the medium. Relatively more growth was observed in all the cases.

5.5 Inoculum Development

Inoculum development is one of the most important variables in the successful completion of a bioprocess run on a laboratory and industrial scale [115]. Inocula for the experiments with *P. anceps* were prepared by growing the cultures on a glucose-yeast extract medium in shaken culture for 4 days homogenizing the mycelium and inoculating 1 ml (0.5 mg dry weight) of this suspension to each flask [85]. This technique for inocula development was found to result in more reproducible results than the technique of inoculating the media with mycelium recovered from agar slants. The inoculum size affects the length of the lag phase in the bioprocesses. A lag phase of 1.5 day, 1.0 day and 0.5 day were observed

Table 9. Different combinations of mineral salts used by various workers

Concentration used (g l^{-1})							
KH$_2$PO$_4$	K$_2$HPO$_4$	CaCl$_2$	MgSO$_4$	ZnSO$_4$	FeSO$_4$	Other additives	Ref.
1.8	5.5	0.1	0.5	0.03	0.1	Yeast extract (0.25)	[76]
0.6	–	–	0.25	0.044	–	–	[84]
–	–	–	–	–	–	Yeast extract (3.0) Malt extract (3.0)	[61, 62 64, 66]
–	1.0	–	0.5	–	0.01	–	[13]
2.0	–	0.4	0.3	–	–	Yeast extract (1.0)	[77]
2.0	–	0.3	0.3	0.0014	0.005	Yeast extract (3.0) Malt extract (3.0)	[63]
1.0	1.0	0.1	0.05	–	–	Yeast extract (1.5)	[86]
1.0	–	0.57	1.02	1.0	1.84	Yeast extract (3.0)	[113]

with inoculum size of 1, 5 and 10% (v/v), respectively. The conversion rate was decreased with an increase of inoculum size above 10%.

In direct bioconversion of cellulose by *N. crassa*, the mould was grown aerobically by inoculating a heavy spore suspension in cellulose medium for 48 h and then the contents were inoculated in the bioconversion flask containing substrate [62]. Increasing inoculum size for a fixed substrate concentration improved the rate of ethanol production by *N. crassa* [63]. In studying ethanol production by *Mucor* and *Fusarium* sp., inocula were prepared by incubating spore suspension for 2 days, harvesting the mycelia by filtration, washing with sterile water and inoculating the bioprocess flasks [13]. Inoculum was prepared in the same medium by transferring a loopful culture from the PDA slants and grown for 48 h; while studying the bioconversion of cellulose to ethanol/acetic acid [66].

5.6 pH

The initial pH values generally employed by various workers were within the range of 5.0–6.0 [13, 62, 64, 66, 76, 77, 80]. Enari and Suihko [33] obtained best results for ethanol production by *F. oxysporum* at pH 5.5 (Table 10), however, a higher acetic acid production was obtained at pH 6.0. Maximum ethanol production by *N. crassa* was found to be in the pH range of 5–6, which is also the optimum pH for cellulase activity [63]. The higher rates were attributed to the faster sugar production from the cellulose by the increased cellulase activity of *N. crassa* at pH 5.0. *Paecilomyces* sp. NF1 was found to possess a wide range of pH

Table 10. Optimum pH and temperature range of ethanol producing fungal species

Organism	Optimum pH	Optimum temperature (°C)	Ref.
Aspergillus niger	–	28	[76]
Fusarium lini	–	28	[76]
F. lycopersici	5.4	30	[13]
F. oxysporum	5.5	30	[33, 37]
F. oxysporum	5.0	30	[66]
F. oxysporum	6	34	[64]
F. oxysporum (*lini*)	5	30	[80]
Monilia sp.	–	26	[61]
Mucor sp.	5.4	30	[13]
Neurospora crassa	5–6	28–37	[65]
Paecilomyces sp.	2.2–7.0	30–37	[86]
Polyporus anceps	6.5–7.0	25	[85]
Rhizopus oryzae	–	35	[84]

optima [86]. This fungus produced the same amount of ethanol when the pH of the initial medium was varied from 2.2–7.0.

Ethanol production by *F. oxysporum* F3 was considerably affected by pH of both aerated and non-aerated cultures [64]. Optimum values were obtained when the pH of the aerated and non-aerated culture of cellulose were 5.5 and 6.0, respectively. It could be due to the changes induced by low pH to systems involved in cellulose hydrolysis, utilization of sugars for bioethanol production, or both [64]. At optimum pH, no insoluble cellulose could be detected in the culture medium. On the other hand, low pH of the aerated culture resulted in low ethanol yield. Adjustment of the initial pH in non-aerated growth to an optimal pH was established to be optimal for both β-glucosidase activity and ethanol production, as a consequence the conversion time resulted to about half [65].

5.7 Temperature

For ethanol production between 25–37 °C (Table 9) various workers found that the effect of temperature was dependant on the strain employed. The optimum temperature for ethanol production from glucose by *N. crassa* was 28–37 °C, whereas maximum ethanol yields from cellulosic substrates were obtained at 37 °C [63]. The bioconversion at 37 °C resulted in more than 90% conversion of cellulose into ethanol within 4 days. Although temperature higher than 37 °C was favourable for cellulase activity but unfavourable for ethanol production. Similarly, *Paecilomyces* sp. was found to have an optimum temperature range of 30–37 °C [86]. The optimum temperature for ethanol production by *F. oxysporum* F3 was 34 °C [64], whereas best performance of *F. oxysporum f.* (*lini*) was found at 30 °C [80].

5.8 Aeration

The rate of utilization of available carbohydrates and eventual conversion to ethanol and other products is significantly influenced by aeration [74, 76]. Ethanol accumulates only under low aeration conditions. In the absence of oxygen, growth is either restricted or fails to occur [34]. Ethanol accumulation under increasing oxygen limitation suggest that this specific oxygen utilization rate is an important factor for the determination of the amount of ethanol produced [121, 122]. With increase in oxygen limitation, growth decreases and ethanol production increases. We have made an attempt to present a comparison of the aeration rate used by various workers and their effect on the ethanol production (Table 11).

In shaken flask studies, the degree of oxygenation is mostly quantified as rev. min^{-1} and culture volume (%). In bioreactor experiments, vol $vol^{-1} min^{-1}$ (vvm) has been evaluated [66, 82]. A series of experiments in a 3 litre bioreactor were carried out to study the effect of aeration on the ethanol production by *F. oxysporum* [33]. Ethanol production appeared to be growth associated and an increased amount of oxygen in the aeration gas was found to increase ethanol productivity. Increase in aeration rate from 0.05 to 0.08 $l^{-1} h^{-1}$ increased the growth but ethanol productivity was not much affected. The best results were obtained using as gas mixture containing 1% oxygen. Ergosterol and Tween-80 did not satisfy the requirement of oxygen. On the other hand, increased amount of oxygen in the aeration gas decreased the production of acetic acid [33]. However, the reverse was true with *Polyporus anceps* [85], when large accumulation of acetic acid was observed with increased aeration. Maximum ethanol yield was obtained with low aeration rate of 10 ml min^{-1}. Interesting oxygen effects have been noted in *Pachysolen tannophilus*. Under aerobic conditions, it produces cell mass, under anaerobic conditions xylitol and under oxygen limitation ethanol is produced [74]. Aeration stimulates sugar utilization in yeasts. Many yeasts related to *Brettanomyces*, *Pichia*, *Endomycopsis* and *Rhodotorula* have been recognised as aerobic yeasts, since no sugar utilization occurs under anaerobic conditions with these organisms [69]. Thus, aeration is an important factor to be optimized in fungal bioprocesses.

6 Ethanol and Sugar Tolerance

End product tolerance is one of the major limiting factors in the bioprocesses [28, 123]. Three major effects of ethanol have been observed viz. cell growth, cell viability and bioprocess productivity. Ethanol tolerance has been found to be strain dependent. The most ethanol tolerant organism known, *Lactobacillus heterohiochii* [124], is able to grow in 20% (w/v) ethanol. Yeast, *Zymomonas* and acetic acid bacteria are able to tolerate 8 to 12% ethanol [28]. The biochemical basis for the ethanol tolerance also vary among the organisms [123]. These include denaturation of glycolytic enzymes, inhibition of sugars, NH_4^+ and amino acid transport, changes in the physico-chemical properties of cell membrane, alteration in optimum temperature of growth and leakage of essential cofactors. Plasma membrane phospholipids also play an important role in the ethanol tolerance mechanism. A considerable amount of work has been done in yeasts and bacteria, however, information available in fungi is scanty.

In the case of *F. oxysporum* (*lini*), the inhibition was apparent only when the ethanol concentration was 15 g l^{-1} [80]. No significant differences were observed

Table 11. Effect of aeration bioconversion of carbohydrates by fungi

Organism	Substrate (g l⁻¹)	Ethanol (g l⁻¹)	Acetate (g l⁻¹)	Vessel volume (ml)	Culture volume (ml)	Rev. min⁻¹	Aeration ml min⁻¹	vvm	Oxygen (%)	Ref.
Aspergillus niger	Glucose (10)	1.1	1.2	250	125	165	—	—	—	[76]
Fusarium oxysporum	Xylose (50)	11.5	1.4	3000	—	—	—	—	0	[33]
	Xylose (50)	22.0	0.9	3000	—	—	—	—	1	[33]
	Xylose (50)	19.0	2.4	3000	—	—	—	—	5	[33]
F. oysporum	Cellulose (10)	2.0	1.7	250	100	100	—	—	—	[66]
	Cellulose (10)	1.5	2.5	5000	3000	400	—	0.04	—	[66]
	Waste pulp (30)	0.1	4.7	5000	3000	400	—	0.04	—	[66]
F. lycopersici	Glucose (160)	2.8	—	250	50	200	—	—	—	[13]
Mucor 101	Glucose (160)	6.6	—	250	50	200	—	—	—	[13]
M. hiemalis	Glucose (10)	2.6	0.3	250	125	165	—	—	—	[76]
Monilia sp.	Cellulose (50)	16	—	250	100	—	—	—	—	[61]
Neurospora crassa	Cellulose (50)	18	—	150	70	—	—	—	—	[62]
Polyporus anceps	Glucose (20)	1.3	1.5	250	40	185	0	—	—	[85]
	Glucose (20)	3.3	1.1	250	40	185	10	—	—	[85]
	Glucose (20)	0.6	1.9	250	40	185	100	—	—	[85]
	Glucose (20)	0.1	2.1	250	40	185	500	—	—	[85]
Rhizopus oryzae	Glucose (150)	6.2	—	—	3000	13	150	—	—	[84]

on the linear growth rate of this mould, when ethanol was added before inoculation or at the mid point of the bioprocess. No growth was observed above $42 \, g \, l^{-1}$ ethanol concentration. Growth inhibition was not observed up to a concentration of 20% xylose. These results indicate high sugar tolerance of fungi.

Ethanol had no inhibitory effect on the rate of glucose conversion up to $45-50 \, g \, l^{-1}$ and xylose conversion up to $35-40 \, g \, l^{-1}$ by *F. oxysporum* [81]. The maximum final concentrations of ethanol obtained were 6.0 and 4.1%, respectively. A glucose concentration of $200 \, g \, l^{-1}$ resulted in $45 \, g \, l^{-1}$ ethanol in 7 days and a conversion of $300 \, g \, l^{-1}$ of glucose solution yielded $43 \, g \, l^{-1}$ ethanol in 14 days by *F. oxysporum* [33]. Xylose concentration of $150-200 \, g \, l^{-1}$ increased the lag phase but did not affect the conversion rate [82]. *N. crassa* does not tolerate substrate concentration higher than $20 \, g \, l^{-1}$ [63].

The rate of ethanol production by *Mucor* was lower when high concentrations of D-xylose ($200 \, g \, l^{-1}$) were used as substrates, indicating that *Mucor* is more sensitive to high concentration of D-xylose than *Fusarium* [13]. At low substrate concentration, reduced yield of ethanol by *Fusarium* F5 was observed after 4 days of incubation, thus, indicating the ability of *Fusarium* to utilise ethanol upon the exhaustion of D-xylose [13]. Similar results were obtained with *F. oxysporum* VTT-D-80134 where sugars were found to be consumed in 2 days and then the organism utilized the ethanol produced previously [33]. Some of the fungal cultures have also been reported which metabolize ethanol into acetic acid [76, 85].

Paecilomyces sp. NF1 was able to produce $73 \, g \, l^{-1}$ ethanol from a xylose concentration of $200 \, g \, l^{-1}$, thus indicating a high tolerance of both xylose and ethanol [86]. Moreover, this mould utilized all the sugars ($30 \, g \, l^{-1}$) from an acid hydrolysate of wheat straw to produce $12.5 \, g \, l^{-1}$ ethanol, showing that it was not affected by toxic substances in hydrolysate like other microorganisms [125, 126].

A few fungal strains have shown high tolerance to sugar, ethanol and toxic substances present in the hydrolysate. Therefore, these strains may be considered potential for ethanol production from a mixture of sugars in the hydrolysate. However, systematic studies on the biochemical mechanism for ethanol tolerance in fungi has not been conducted so far. This is of great interest since improved strains with increased tolerance will be more productive.

7 Epilogue

Biological production of ethanol is a potential route to partially replace oil and chemical feedstock. The overall cost of producing chemicals depends on the cost of raw materials, cost of plant and running cost [28]. Lignocellulosic substrates are sources of cheap raw materials and potentially have important features from energetic and environmental viewpoints [113]. However, the difficulty in economically converting these components is responsible for the incomplete success in developing practical process for bioethanol production. *Saccharomyces* and *Zymomonas* metabolize a limited range of substrates, while some fungal strains e.g. *Fusarium*, *Monilia*, *Paecilomyces* and *Neurospora* possess the ability of metabolizing a wide range of substrates including cellulose [62–66, 68]. The economic importance of obtaining ethanol from natural cellulosic materials mostly depends on the conversion of both hexoses and pentoses present in the hydrolysate. Since some raw materials contain a high pentose content, considerable interest has been focussed on microbial species that can utilize these sugars. However, the bioconversion of pentose sugars is slower than that of hexose. This limitation can be amended by strain selection using contemporary recombinant DNA technology, but more basic knowledge about the physiology and biochemistry of the pentose bioconversion is necessary.

It is apparent from the foregoing that although no available fungal strain is completely satisfactory for the conversion process of cellulosic biomass to ethanol, the ability of fungi for biological conversion of cellulosic materials, starches and pentoses has a direct bearing on the economics of producing chemical feedstock. Efficient utilization of the hemicellulose and cellulose hydrolysate by some fungal cultures also indicate resistance from toxic inhibitors present in the natural substances [13, 86]. However, the most limiting factor is the slow conversion rate. Single stage direct conversion of low cost biopolymeric substrates for the production of value added chemicals as well as polysaccharase enzymes possess several advantages. The most important being the use of single bioreactor which simplifies the process and reduce the capital cost. A further advantage is increase in the overall rate of conversion, since the intermediate products are removed as they are formed. Thus, feedback inhibition and catabolite repression of poly-merases are relieved. However, improvement in the bioconversion rate which is the most limiting factor with fungal cultures is necessary. Nevertheless, the simplicity inherent in the direct biological conversion process tends to promote further research.

8 References

1. Lipinski CV (1981) Science 212: 1466
2. Rosenberg SL (1980) Enz Microb Technol 2: 185
3. Klausmier WH (1983) Biotechnol Bioeng Symp 13: 81
4. Soltes EJ, Lin SCK (1983) Biotechnol Bioeng Symp 13: 53
5. Reed TB, Milne T, Diebold J, Derosier R (1981) Biotechnol Bioeng Symp 11: 137
6. Zeikus JG (1980) Ann Rev Microbiol 34: 423
7. Keim CR (1983) Enz Microb Technol 5: 103
8. Ladisch MR (1979) Proc Biochem 14: 21
9. Bullock JD (1979) Industrial alcohol. In: Bull AT, Ellwood, DC, Ratledge C (eds) Microbial Technology: Current Status and Future Prospects, Cambridge University Press, New York, p 309
10. Tsao GT (1978) Proc Biochem 13: 12
11 Stewart GG, Panchal CJ, Russel I, Sills AM (1984) CRC Crit Rev Biotechnol 1: 161
12. Schneider H, Wang PY, Chan YK, Maleszka R (1981) Biotechnol Lett 3: 89
13. Ueng PP, Gong CS (1982) Enz Microb Technol 4: 169
14. Jeffries TW (1984) Trends Biotechnol 3: 208
15. Wang DIC, Averingos GC, Biocic I, Wang SD, Fang HW (1983) Phil Trans Roy Soc London 300: 323
16. Jones RP, Pamment N, Greenfield PF (1981) Proc Biochem 16: 42

17. Hoffman H, Kuhlman W, Meyer H-D, Schügerl K. (1985) J Memb Sci 22: 235
18. Frick C, Schügerl K (1986) Appl Microbiol Biotechnol 25: 186
19. Zeikus JG, Ben-Bassat A, Ng TK, Lamed RJ (1981) Thermophilic ethanol fermentation. In: Hollaender A (ed) Trends in the Biology of Fermentation for Fuels and Chemicals, Plenum Press, New York, p 44
20. Rogers PL, Lee KJ, Tribe DE (1979) Biotechnol Lett 1: 165
21. Doelle HW (1982) Eur J Appl Microbiol Biotechnol 15: 20
22. Cromie S, Doelle HW (1981) Eur J Appl Microbiol Biotechnol 11: 16
23. Cromie S, Doelle HW (1982) Eur. J Appl Microbiol Biotechnol 14: 69
24. Dawes EA, Ribbons DW, Rees DA (1966) Biochem J 98: 804
25. Lee JH, Williamson D, Rogers PL (1981) Biotechnol Lett 3: 291
26. Karsh T, Stahl U, Esser K (1983) Eur J Appl Microbiol Biotechnol 18: 383
27. Rose AH, Harrison JS (1971) The Yeasts, Vol 2 & 3, Academic Press, New York
28. Lowitt RW, Kim BH, Shen G-J, Zeikus JG (1988) CRC Crit Rev Biotechnol 7: 107
29. Marwaha SS, Kennedy JF (1984) Enz Microb Technol 6: 18
30. Frelot D, Moulin G, Galzy P (1982) Biotechnol Lett 4: 411
31. Margarits A, Bajpai P, Cannel E (1981) Biotechnol Lett 3: 595
32. O'Leary VS, Sutton C, Bencivengo M, Sullivan B, Holsinger VA (1977) Biotechnol Bioeng 19: 1689
33. Enari T-M, Suihko M-L (1984) CRC Crit Rev Biotechnol 1: 229
34. Schneider H (1989) CRC Crit Rev Biotechnol 9: 1
35. Pelvako EA, Czeban ME (1935) Mikrobiologiya 4: 86
36. Wang PY, Shopsis C, Schneider H (1980) Biochem Biophys Res Comm 94: 248
37. Wang PY, Johnson BF, Schneider H (1980) Biotechnol Lett 3: 273
38. Silinger PJ, Botahst RJ, van Cauwenberge JE, Kurtzman CP (1982) Biotechnol Bioeng 24: 371
39. Sreenath HK, Chapman TW, Jeffries TW (1986) Appl Microbiol Biotechnol 24: 294
40. Schvester P, Robinson CW, Moo-Young M (1983) Biotechnol Bioeng Symp 13: 131
41. Gong CS, Chen LF, Flickinger MC, Chiang LC, Tsao GT (1981) Appl Environ Microbiol 41: 430
42. Karczewska H (1959) CR Trav Lab Carlsberg 31: 251
43. Schneider H (1982) The production of ethanol from D-xylose by yeasts. In: Proc. Roy Soc Can Int Symp Ethanol From Biomass, Winnipeg, p 508
44. Maleszka R, Veliky IA, Schneider H (1981) Biotechnol Lett 3: 415
45. Kosaric N, Ng DCM, Russel I, Stewart GG (1982) Adv Appl Microbiol 24: 147
46. Spangler DJ, Emert GH (1986) Biotechnol Bioeng 28: 115
47. Ooshima H, Ishtani Y, Harano Y (1985) Biotechnol Bioeng 27: 389
48. Spindler DD, Wyman, CE, Grohmann K (1989) Biotechnol Bioeng 34: 189
49. Deshpande V, Sivaraman H, Rao M (1983) Biotechnol Bioeng 25: 1679
50. Saddler JN, Hogan C, Chan MKH, Louis-Seize G (1982) Can J Microbiol 28: 1311
51. Ghosh P, Pamment NB, Martin WRB (1982) Enz Microb Technol 4: 425
52. Klyosov AA, Consultant V (1986) Appl Biochem Biotechnol 12: 249
53. Szczodrak J, Targonski Z (1988) Biotechnol Bioeng 31: 300
54. Chen S, Wayman M (1989) Proc Biochem 24: 204
55. Cooney CL, Wang DIC, Wang S-D, Gordon J, Jiminez M (1978) Biotechnol Bioeng Symp 8: 103
56. Florenzano G, Poulain M, Goma G (1984) Biomass 4: 295
57. Saddler JN, Chan MKH, Seize GL, (1981) Biotechnol Lett 3: 321
58. Ng TK, Bassat AB, Zeikus JG (1981) Appl Environ Microbiol 41: 1337
59. Hynn HH, Shen GJ, Zeikus JG (1985) J Bacteriol 164: 1153
60. Averinos GC, Fang HY, Biocic I, Wang DIC (1981) Adv Biotechnol 2: 119
61. Gong CS, Mann CM, Tsao GT (1981) Biotechnol Lett 3: 77
62. Rao M, Deshpande V, Keskar S, Srinivasan MC (1983) Enz Microb Technol 5: 133
63. Deshpande V, Keskar S, Mishra C, Rao M (1986) Enz Microb Technol 8: 149
64. Christakopoulos P, Macris BJ, Kekos D (1989) Enz Microb Technol 11: 236
65. Christakopoulos P, Macris BJ, Kekos D (1990) Appl Microbiol Biotechnol 33: 18

66. Kumar PKR, Singh A, Schügerl K (1991) Appl Microbiol Biotechnol 34: 570–572
67. Singh A (1988) Biochemical studies on utilization of agricultural wastes using cellulolytic fungi. Thesis, N.D. University of Agriculture and Technology, India
68. Marsden WL, Gray PP (1986) CRC Crit Rev Biotechnol 3: 235
69. Jeffries TW (1983) Adv Biochem Eng Biotechnol 27: 2
70. Rogers PL, Lee KJ, Tribe DE (1980) Adv Biochem Eng Biotechnol 23: 37
71. Swings J, De Ley I (1977) Bacteriol Rev 41: 1
72. Zeikus JG (1979) Enz Microb Technol 1: 243
73. Schneider H, Maleszka R, Neirinck L, Veliky IA, Wang PY, Chan YK (1983) Adv Biochem Eng/Biotechnol 25: 57
74. Skoog K, Hahn-Hägerdal B (1988) Enz Microb Technol 10: 66
75. Cochrane VW (1958) Physiology of Fungi, John Wiley and Sons Inc., London, p 214
76. Perlman D (1950) Am J Bot 37: 237
77. Suihko M-L, Enari T-M (1981) Biotechnol Lett 3: 723
78. Batter TR, Wilke CR (1977) A study of the fermentation of xylose to ethanol by *Fusarium oxysporum* Energy and environment division, Lawrence Berkley Laboratory, University of California, Berkley, 1977
79. Antonopoulos AA Wene EG (1984) Phytopathol 74: 1268
80. Rosenberg Sl, Batter TR, Blanch HW, Wilke CR (1981) AIChE Symp Ser 77, 107
81. Suihko M-L (1983) Biotechnol Lett 5: 721
82. Linko M, Viikari L, Suihko ML (1984) Biotechnol Adv 2: 233
83. Waksman SA, Foster JW (1938) J Agri Res 57: 873
84. Ward GE, Lockwood LB, Tabenkin B, Well PA (1938) Ind Eng Chem 30: 1233
85. Perlman D (1949) Am J Bot 36: 180
86. Wu JF, Lastick SM, Updegraff DM (1986) Nature 321, 887
87. Anderson AK (1924) Minn St Plant Sci 5: 237
88. Letcher H, Willmann JJ (1926) Phytopathol 16: 941
89. Antonopoulos AA (1979) *Fusarium* species: Their potential for transforming biomass to ethanol. Energy and Environmental Systems Division, Argonne National Laboratory, Argonne, Illinois 60439
90. Fiechter A, Furhmann GF, Käppeli O (1981) Adv Microb Physiol 35: 454
91. Chiang C, Knight SG (1959) Biochim Biophys Acta 35: 454
92. Chiang C, Knight SG (1960) Nature 118: 79
93. Suhiko M-L, Suomalainen I Enari T-M (1983) Biotechnol Lett 5: 525
94. Tomoyeda M, Horitsn H (1964) Agric Biol Res 28: 139
95. Horitsu H (1973) Proc 3rd Int Spec Symp Yeasts, Helsinki, p 136
96. Hofer M, Betz A, Kotyk A (1971) Biochem Biophys Acta 252: 1
97. Zanek J, Kalecikova B, Kuniak L, Kucas S, Kockovakratochvilova A (1986) Acta Aliment 15: 111
98. Chiang C, Knight SG (1960) Biochem Biophys Res Comm 3: 554
99. Peixoto BG, Viega LA (1974) Arq Biol Technol 17: 87
100. Gibbs M, Cochrane VW, Paege LM, Wolin H (1951) Arch Biochem Biophys 33: 346
101. Coleman RJ, Nord FF (1951) Arch Biochem Biophys 33: 346
102. Halliwell G, (1961) Biochem J 79: 185
103. Halliwell G, Griffin M (1973) J Gen Microbiol 216: 211
104. Alexander M (1977) Introduction to Soil Microbiology, Wiley Eastern Ltd., New Delhi, p 386
105. Pathak AN, Ghosh TK (1973) Proc Biochem 8: 35
106. Mandels M, Weber J, Adv Biochem Ser 95: 391
107. Shewale JG, Sadana JC (1978) Can J Microbiol 24: 1204
108. Singh A, Abidi AB, Darmwal NS, Agrawal AK (1988) Bioconversion of agricultural lignocellulosic residues for the production of protein and cellulase in solid state culture. "Application of Biotechnology for Agriculture and Rural Development", NIRD, Hyderabad, p 68
109. Tangnu SK, Blanch HW, Wilke CR (1981) Biotechnol Bioeng 23: 1837
110. Ishaque M, Kluepfel D (1981) Biotechnol Lett 3: 481

111. Viikari L, Klemola M, Linko M (1978) Production of xylanases. Proc. Bioconversion in Food Technology, Helsinki, p 115
112. Mishra C, Keskars, Rao M (1984) Appl Environ Microbiol 48, 224
113. Lynd LR (1989) Adv Biochem Eng/Biotechnol 38: 1
114. Chiang LC, Hsiao HY Flickinger MC, Chen LF, Tsao GT (1982) Enz Microb Technol 4: 93
115. Kumar PKR, Lonsane BK (1989) Adv Appl Microbiol 34: 29
116. Schneider H (1989) Conversion of hemicelluloses into ethanol. In: Chahal DS (ed) Food, Feed and Fuel from Biomass, Oxford and I.B.H. Publishing C., New Delhi
117. Silinger PJ, Bothast RJ, Okos MR, Ladisch MR (1985) Biotechnol Lett 7: 431
118. Jeffries, TW, Fady JH, Lightfoot, FN (1985) Biotechnol Lett 7: 171
119. du Preez JC, Bosch M, Prior BA (1986) Enz Microb Technol 8: 360
120. Sale A (1967) Regulation of carbohydrate transport and metabolism in yeast. In: Mills AK, Krebs H (Eds) Aspects of Yeast Metabolism, F. A. Davies & Co., Philadelphia, p 45
121. Chung IS, Lee YY (1986) Enz Microb Technol 8: 503
122. Mahmourides G, Lee H, Maki N, Schneider H (1985) Bio/Technology 3: 59
123. D'Amore T, Stewart GG (1987) Enz Microb Technol 9: 322
124. Takagawa T, Wmezu M (1979) J Gen Appl Microbiol 25: 41
125. Lee YY and McCaskey TA (1983) Tappi J 66: 102
126. Hueting S, Tempest DW (1977) Arch Microbiol 115: 73
127. Singh, A, Kumar, PKR, Schügerl, K (1991) in preparation

The Enzymes from Extreme Thermophiles: Bacterial Sources, Thermostabilities and Industrial Relevance

T. Coolbear*, R. M. Daniel and H. W. Morgan
Thermophile Research Unit, University of Waikato, Private Bag, Hamilton, New Zealand

Table of Contents

This review on enzymes from extreme thermophiles (optimum growth temperature > 65 °C) concentrates on their characteristics, especially thermostabilities, and their commercial applicability. The enzymes are considered in general terms first, with comments on denaturation, stabilization and industrial processes. Discussion of the enzymes subsequently proceeds in order of their E.C. classification: oxidoreductases, transferases, hydrolases, lyases, isomerases and ligases. The ramifications of cloned enzymes from extreme thermophiles are also discussed.

---

* Corresponding author; Present address: New Zealand Dairy Research Institute, Private Bag, Palmerston North, New Zealand

Advances in Biochemical Engineering/
Biotechnology, Vol. 45
Managing Editor: A. Fiechter
© Springer-Verlag Berlin Heidelberg 1992

1 Introduction

The purpose of this review is to provide an overview of enzymes from extreme thermophiles with particular emphasis on thermostability characteristics, and with some reference to the commercial applicability of the enzymes. An extreme thermophile is defined here as an organism with a growth optimum of 65 °C or higher. This limits the scope of the review to bacteria, since no extremely thermophilic fungi or algae have been isolated, and excludes much of the work carried out on *Bacillus stearothermophilus*.

Bacteria which have been identified as extreme thermophiles represent both the eubacterial and the archaebacterial kingdoms [1]. However, with the exception of *Thermotoga* species, all known organisms growing optimally above 75 °C are archaebacteria. The highest optimum growth temperature recorded for a living organism is 105 °C for the archaebacteria *Pyrodictium brockii* and *P. occultum*. These organisms were found to be capable of growth at 110 °C [2] this also being the upper growth temperature limit for "*Methanopyrus*" [3] and the archaebacterial strain ES4 [4]. These bacteria, together with *Hyperthermus butylicus* [5], are the most thermophilic organisms known to date.

The archaebacteria differ from eubacteria in that they lack muramic acid and therefore a peptidoglycan cell wall [6, 7], their membrane lipids are based on ether linkages, rather than ester linkages [8–11] and they differ in intermediary metabolism and coenzyme complement [12, 13]. The differences between archaebacteria and eubacteria imply that the former are a source of unique enzymes, but so far only a few of these have been studied (see below).

The most extremely thermophilic eubacteria are the two *Thermotoga* species, *T. maritima* [14, 15] and *T. neopolitana* [16]. *Thermotoga* strains are unusual eubacteria having unique properties in cell wall structure and lipid composition [14], and, using RNA sequencing analysis to construct phylogenetic trees, they have been shown to represent the deepest branch in the eubacterial line of descent [17]. This is further evidence for a thermophilic origin for eubacteria.

Taxonomic studies on extremely thermophilic eubacteria are limited [18–24]. Many extreme thermophiles currently designated as belonging to a particular species of *Bacillus* or *Thermus* may be reclassified when their taxonomy is better understood.

2 Thermostable Enzymes — General Considerations

2.1 Denaturation and Degradation

Enzyme inactivation by heat occurs by conformational unfolding (denaturation) and by irreversible reactions involving the cleavage of covalent bonds (degradation). In denaturation the disruption of the non-covalent forces stabilizing protein structure causes a reversible loss of conformation. This, however, is usually followed by an aggregation step which is effectively irreversible. At high temperatures

(> 100 °C) irreversible degradative processes such as peptide bond hydrolysis and deamidation become important [25–28].

On theoretical grounds, there seems no reason why conformational integrity should not be maintained well above temperatures at which all enzymes are known to lose activity rapidly, and degradation seems likely to set the upper limit for protein stability. However, at least some reactions responsible for degradation are dependent on the flexibility of the polypeptide chain backbone, and on the arrangement of amino acids. Thus protein may be able to evolve to minimise degradation. Recent results in our laboratory show that under the right conditions some enzymes can have half-lives of over 10 min at 125 °C.

2.2 Enzyme Stability

A number of factors may affect enzyme stability in vivo. Certain enzymes from thermophiles have been shown to be less stable in the pure state or in cell-free extracts than in vivo, integrity being enhanced by factors in the cellular environment [29], such as high protein concentration, substrates, prosthetic groups and other molecules [30–34]. The fact remains, however, that thermophilic enzymes as a class are inherently more thermostable than their mesophilic counterparts, and that this stability is primarily intrinsic rather than being determined by extrinsic factors.

Mechanisms of protein stability have been discussed in detail elsewhere [35–39] but some general factors warrant mention here. Point mutations [40–42] and studies on differences in amino acid sequences of related proteins [43–44] have shown that changes in overall structure need only be very small for thermostability to be dramatically altered. Protein stability is the resultant of large destabilizing forces, of the order of $1000 \, kJ \, mol^{-1}$, due to peptide chain entropy and only slightly larger stabilizing forces due to the sum of electrostatic and hydrophobic interactions and hydrogen bonds. The net free energy of stabilization of proteins is generally of the order of $30 \, kJ \, mol^{-1}$. Thus very small percentage changes in the stabilizing forces can have dramatic effects on overall stability. Even a single additional interaction can contribute to the order of $10 \, kJ \, mol^{-1}$. Thus, only a few additional salt bridges, hydrophobic interactions or hydrogen bonds can easily account for the increase in free energy of stabilization required to elevate the relative stability of an enzyme by several degrees centigrade [see 38]. At temperatures exceeding 100 °C hydrophobic bonding may not be a primary effector of protein folding or stability. At such temperatures an increased number of internal salt bridges may be major stabilizing factors.

2.3 Stabilization of Enzymes

With the lack of thermostability of mesophilic enzymes cited as a major drawback to the use of such enzymes in industry, efforts have been made not only to stabilize mesophilic enzymes by protein and genetic engineering or by chemical means, but also to produce totally artificial catalysts, the active centres of which are modelled on the active site of the appropriate enzyme. The latter case is illustrated by the cyclodextrin based molecule "β-benzyme" which mimics the activity of chymotrypsin but is inherently more stable to extremes of heat, pH and ionic strength [45, 46]. The possibilities of constructing artificial, catalytically active polypeptides with appropriate tertiary structure for stability have been recently reviewed by Mutter [47]. The attempts made using site-directed mutagenesis to stabilize mesophilic proteins against denaturation (such as those of Ahern et al. [48]) have, however, met with generally limited success mainly because of difficulty in establishing target areas for mutation. On the other hand, protein engineers have had substantial success in modifying the specificity of enzymes by manipulation of the active site amino acids (a more specific target than amino acids involved in stabilization), and tailoring the active sites of naturally thermostable enzymes to meet industrial requirements would seem a more productive goal. It is worth noting that the factors which are responsible for stabilization of enzymes from extreme thermophiles also confer resistance to other denaturants such as chaotrophic agents, ionic strength, organic solvents and detergents and, indeed, to proteolysis [49–51], possibly by virtue of increased conformational rigidity. Vihinen [52] examined the flexibility of proteins with reference to thermostability and found that thermostable proteins required higher temperatures than their thermolabile counterparts before exhibiting a comparable degree of flexibility, reflecting their catalytic temperature preferences. An upper limit may be placed on the usefulness of enzyme engineering by the competing demands of resistance to unfolding (for thermostability) with the need for flexibility for catalytic efficiency. Successes in increasing thermostability by mutation have generally led to decreased catalytic efficiency [see e.g. 53].

Experiments with lactate dehydrogenases from mesophilic and thermophilic bacilli have led to the production of a hybrid gene encoding for approximately the first one hundred amino acids of the *Bacillus stearothermophilus* enzyme and the second two hundred amino acids from the *B. megaterium* enzyme, the resulting enzyme having intermediate stability but no activation response to fructose-1,6-bisphosphate [54]. Expression in thermophilic clones of genetic information derived from a mesophilic organism has also been used in attempts to improve thermostability of a mesophilic enzyme, as in the case of kanamycin nucleotidyltransferase from *Escherichia coli* [55]. This enzyme was expressed in *Bacillus stearothermophilus* grown at 63 °C or above and the most stable mutant was shown to have a half-life at 65 °C of approximately 15 min, compared to less than 2 min at 55 °C for the wild-type enzyme. Two amino acid substitutions were found in the mutant enzyme, a tyrosine replacing an asparagine and a lysine replacing a threonine (only the former mutation was found in less stable enzyme variants).

Chemical modifications of mesophilic enzymes have met with more success than protein engineering. The strategies for stabilization have been reviewed [e.g. 39, 56] and include immobilization and hydrophilization. Chemical modification of trypsin and chymotrypsin gives artificial hydrophilization of non-polar areas of the protein surface resulting in marked thermostabilization and little loss of enzyme activity [57–59]. This phenomenon has also been demonstrated for *Bacillus stearothermophilus* lactate dehydrogenase by Wigley et al. [60] who used site-directed mutagenesis to replace a surface leucine residue with arginine. Other specific studies on thermostability invoke the roles of surface arginyl residues [61, 62] or a high arginine to lysine ratio [63] (this in contrast to later studies [64, 65]) and high proline content [66] in stabilization. Also of relevance are the results of Matthews et al. [67] with bacteriophage T4 lysozyme, whereby replacement of certain amino acids with proline and substitution of glycine with amino acids of less conformational flexibility lead to decreased second order thermal inactivation rate constants. Imanaka et al. [68] and Querol and Parrilla [69] have suggested some guidelines for increasing stability by site directed mutagenesis, but there is little information available to judge their applicability.

Recently the glyceraldehyde-3-phosphate dehydrogenases from mesophilic and thermophilic archaebacteria have been shown to be 70% homologous [64]. Homology to the corresponding enzyme from eubacterial and eukaryotic organisms is low, but Fabry et al. [64] argue that within secondary structure configurations amino acid exchanges conferring thermostability could be distinguished, notably a preference for isoleucine and a low glycine content, and a low arginine to lysine ratio. These results are mainly (but not completely) consistent with the generalised principles for amino acid exchanges that have been suggested recently [66]. Data from studies on the glyceraldehyde-3-phosphate dehydrogenase from *Thermotoga maritima*, however, suggested that thermostabilization of the enzyme from this eubacterium could not be attributed to amino acid exchanges [70, 71].

Overall, the subtleties of protein structure could preclude the development of widely applicable strategies for engineering greater stability into enzymes. It is likely that appropriate manipulations will differ for each enzyme or enzyme structure type and this presupposes detailed knowledge of the molecules.

Oligomeric enzymes are mooted as being additionally heat stabilized by inter-subunit interaction, as in the tetrameric D-glyceraldehyde-3-phosphate dehydrogenase from *B. stearothermophilus* [72]. The resistance to acid-mediated dissociation of malate dehydrogenase of *"Thermus flavus"* was ascribed to the enhanced subunit interactions implicated in overall heat stability [73]. Chemically catalysed intramolecular cross-linking has been used to stabilize monomeric enzymes [74] and a similar process for inter-subunit crosslinking of polymeric enzymes using diacids (e.g. succinic acid) is also effective [75]. This artifical cross linking has been proposed as parallelling salt bridge stabilization in thermophilic enzymes [76].

A further mechanism of thermostability in thermophilic enzymes is reversible aggregation, a phenomenon most recently reported for glutamine synthetase from *"B. caldolyticus"* [77]. Such aggregation has been proposed as the reason for the elevated thermostability of this enzyme from both *"B. caldolyticus"* and *B. stearothermophilus* in concentrated solutions [30, 78]. A similar proposal has been made for the glutamate dehydrogenase from *Sulfolobus* (Scandurra R, pers. comm.).

2.4 Practical Aspects

Clear parameters need to be defined with regard to activity at high temperatures and resistance to inactivation by heat in order to compare enzymes. A plot of enzyme activity versus temperature is a result of two effects, the acceleration of catalysis due to increased temperature and losses due to heat denaturation. The position of the peak is dependent on the duration of the assay and without this information the all too common term "optimum temperature" is useless, and even then gives little information concerning stability. Similarly the phrase "the enzyme was stable at 75 °C" begs the question as to length of time. These descriptions have been noted repeatedly in papers collated for this review and in some instances the extraction of comparative data has not been possible. From a practical point of view, the terms "optimum temperature" ("maximum temperature"!) and "stable" should be fully defined before being used.

Many commonly used buffers have large temperature coefficients. Even if plots of initial activity versus temperature under uniform conditions can be obtained they must be assessed with caution, and possible temperature-induced changes in substrate conformation, enzyme pH optimum, and Km taken into account. Estimations of activation energies from true Arrhenius plots and interpretations of any breaks in the slopes require careful consideration [see 79–82]. Caution must also be exercised when interpreting kinetic data for enzymes from extreme thermophiles if assay temperatures are used which are below the source temperature or laboratory growth temperatures. While accurate enzyme assays (especially continuous ones) present some technical problems above 75 °C, low temperature data may differ significantly from that obtained in the physiological range. For example, the glutamate dehydrogenase from *Thermococcus* strain AN1 (which has an optimum growth temperature of 80 °C) exhibits a Km for glutamate of 2 mM at 60 °C while at 80 °C the Km is 11 mM (Hudson R C, Daniel R M, unpublished observations).

The most convenient and comparable measurement of enzyme stability is the half-life at a given temperature (other incubation conditions being defined), i.e. the time period over which half the initial activity of an enzyme is lost. Additional information on the effect of temperature on activity can be important, for instance the shape of a plot of $\log_{10}$ of percent activity remaining versus time of incubation can indicate whether irreversible thermal denaturation is the sole cause of loss of activity or whether additional factors are involved, such as autolysis, renaturation, or substrate effects.

An additional complicating factor is the phenomenon of heat activation of an enzyme over the first few minutes of stability studies [e.g. 83, 84 and Whittaker J M, Daniel R M: unpublished observations]. This has been associated with metal ion equilibration in the active site, reversal of cold induced dissociation of subunits of cold labile enzymes and release of conformational constraints. Activation has also been observed on exposure of thermophilic enzymes to denaturing agents (Rossi M, pers. comm.).

3 Advantages of High Temperature in Bio-Industrial Processes

The potential advantages of using thermophilic enzymes (and organisms) in bio-industrial processes have been considered in numerous recent reviews [85–92] and only a brief discussion will be presented here.

The benefits of using enzymes as catalysts in industrial processes lie in their specificity and efficiency, giving fewer side products, less toxic waste and reduced handling problems. Their main disadvantages reside in stability problems and high cost. The latter is partially consequent on the former since the frequency of replacement of an enzyme in a reactor (and therefore total production costs) is stability-dependent. Production cost itself is generally high due to low yields of enzyme per unit biomass, the expense of extraction and purification and losses of activity incurred during purification, storage and handling — stability of the enzyme again being a factor. As has been discussed already the use of thermophilic enzymes reduces stability problems and in so doing alleviates some of the expense of production and replacement in a reactor.

For many of the extreme thermophiles defined media are not yet available, enzyme yields are low and some of the archaebacteria are very fastidious with low cell yields being obtained. Cloning and expression of the enzymes in *E. coli* (for instance) can obviate these problems and notable advances in this area have occurred very recently (see Sect. 6).

The stability of enzymes from thermophiles should lead to higher recoveries at ambient temperatures than is possible for mesophiles, and even at the higher temperatures possible losses may be offset by more efficient recovery processes resulting from the physical characteristics of the hot liquids. The low activity of extremely thermophilic enzymes at ambient temperatures eases handling and storage problems and the comparative molecular inflexibility that results in this inactivity at lower temperatures has been suggested to lower the immune response to such proteins [93], thus reducing potential health risks. These risks are also reduced in as much as the organisms themselves are not pathogenic (being unable to grow at body temperature) and the culture temperatures of the extreme thermophiles would preclude growth of pathogenic contaminants. This does not necessarily mean that hazards are eliminated since toxin production by extreme thermophiles is possible, although there is no evidence for this to date.

4 Thermophilic Enzymes

The enzymes will be discussed in class order, thermostability data being presented in Table 1.

4.1 Oxidoreductases

The potential applications of oxidoreductases in industry are numerous. Alcohol dehydrogenases are useful in stereospecific synthesis and production of high-cost compounds such as cyclic ethers (see below), and have also been used in the production of flavour aldehydes e.g. geraniol [94]. With mesophilic enzymes, however, limitations due to narrow specificity, instability to heat and organic solvents and loss of activity on immobilization have been incurred. The use of these enzymes from thermophilic sources in immobilized and continuous reactor systems (including biosensors) has also been proposed for the regeneration of

Table 1. Thermostabilities of enzymes from extreme thermophiles. Enzymes are listed under class headings in order of discussion in the text

Enzyme†	Organism*	Thermostability data**	Comments**	Ref.
OXIDOREDUCTASES				
ADH	*S. solfataricus*	5 h, 70 °C		100
ADH (2°)	*Th. ethanolicus*	st. 2 h, 70 °C	Arrhenius plot data suggest 1° ADH less stable	101, 102
ADH (2°)	*Th. brockii*	>20 min, 90 °C		103
LDH	*Thermus* sp.	~4 h, 95 °C	F-1,6-bisP stabilizes: no loss 1 h, 95 °C	105
LDH	*Thermotoga maritima*	~2 min, 90 °C	NAD^+, F-1,6-bisP stabilizes: 2–5 h, 90 °C	114
MDH	*"T. flavus"*	st. 1 h, 90 °C		73, 117 –120
MDH	*T. aquaticus*	20 min, 95 °C		121
MDH	*T.* sp (Iceland)	20–45 min, 95 °C; 4–6 min, 100 °C	Cf *B. stearothermophilus*: 5 min, 75 °C	122
MDH	*Tp. acidophilum*	10 min, 77 °C		124
MDH	*S. acidocaldarius*	<15 min, 90 °C	More stable in crude form	125
MDH	*M. fervidus*	8 min, 90 °C	16 h, 90 °C with cyclic 2,3 di-phosphoglycerate	128
Malic Enzyme	*S. solfataricus*	5 h, 85 °C	Stable in 50% dimethylformamide 24 h, 25 °C	132, 133
Isocitrate DH	*T. aquaticus*	1 h, 88 °C; 32 min, 90 °C	Buffer dependent: Tris < Hepes < PO_4^{2-} · Isocitrate stabilizes	135
Isocitrate DH	*T. thermophilus* HB8	st. 10 min, 70 °C	75% loss 10 min, 90 °C	136
isoPM DH	*T. thermophilus* HB8	~5.5 min, 86 °C	Stabilized by 2 M KCl	137
Alanine DH	*T. thermophilus*	7.5 h, 80 °C; 10 min, 85 °C		138
GDH	*S. solfataricus*	45 h, 70 °C		139
G-3-P DH	*B. stearothermophilus*	20 min, 75 °C		140
G-3-P DH	*T. aquaticus*	~20 min, 100 °C		141
G-3-P DH	*Thermotoga maritima*	>2 h, 100 °C	Optimum stability at pH 6	71

Enzyme	Organism	Stability	Notes	Ref.
G-3-P DH	M. fervidus	30 min, 84 °C; 1.5 min, 90 °C	Stabilized by NADP, 3.5 h, 90 °C with cyclic diphosphoglycerate	128, 142
G-3-P DH	Tpr. tenax	20 min, 100 °C	NAD$^+$ dependent	143
G-3-P DH	Tpr. tenax	35 min, 100 °C	NADH dependent	143
G-3-P DH	P. woesei	44 min, 100 °C; 4.5 min, 107 °C	Stable 30 min, 104 °C with potassium citrate	147
NADH-DH	S. acidocaldarius	30% loss: 15 min, 100 °C		148
NADH-DH	B. caldovelox	40% loss: 30 min, 60 °C		149
NADH-DH	T. aquaticus	35 min, 95 °C	Membrane lipids included	83, 150
Hydrogenase	C. hydrogenophilum	4.5 min, 100 °C		153
Hydrogenase	Mc. jannaschii	37 min, 80 °C; 6.2 min 85 °C	F_{420} non reactive. 1 min, 100 °C in 50% glycerol	154
Hydrogenase	P. furiosus	9.2 min, 80 °C; 5.2 min, 85 °C; 21 h, 80 °C; 5 min, 105 °C	F_{420} reactive. Destabilized under aerobic conditions	155
TRANSFERASES				
DNA polymerase	T. ruber	10% loss: 2 h, 70 °C		159
DNA polymerase	"T. flavus"	2 h, 70 °C	T. aquaticus enzyme needs similar conditions for max. activity	160, 161
DNA polymerase	T. thermophilus HB8	2 min/45 min/5 min, 90 °C	For A, B & C forms respectively. A & C biphasic	162
DNA polymerase	Thermotoga sp.	3 min, 95 °C	Triton X-100 and mannitol stabilize additively	164
DNA polymerase	S. solfataricus	35 min, 85 °C; 6 min, 90 °C		167
DNA polymerase	S. acidocaldarius	30 min, 85 °C		168, 169
DNA-RNA poly.	M. fervidus	84% loss: 2 min, 80 °C	Not stabilized by BSA, DNA, glycerol, reducing agents	170
DNA-RNA poly.	Tpr. tenax	>2 h, 100 °C		171
DNA-RNA poly.	S. acidocaldarius	10 min, 83.5 °C		(171)
DNA-RNA poly.	D. mucosus	23 min, 95 °C		
tRNA MeTase	T. flavus	st. 1 h, 60 °C		176
tRNA MeTase	T. thermophilus HB27	4 h, 70 °C; 16 min, 90 °C	100% loss in 5 min at 90 °C	155–157
Propylamino Tase	S. solfataricus	st. 1 h, 100 °C		181

(Continued page 10)

Table 1. Continued

Enzyme†	Organism*	Thermostability data**	Comments**	Ref.
Asp amino Tase	S. solfataricus	2 h, 100 °C	pridoxamine 5'-phosphate, BSA, 2-oxoglutarate present	182, 183
Phe amino Tase	T. aquaticus YT1	25 min, 70 °C		185
Phe amino Tase	B. stearothermophilus	25% loss: 25 min, 70 °C		
Phe amino Tase	B. stearothermophilus	25% loss: 25 min, 75 °C		
PF Kinase	T. aquaticus YT1	st. 24 h, 80 °C		186
PF Kinase	F. thermophilum	5 h, 90 °C		187
Pyruvate kinase	T. thermophilus	35 min, 90 °C	Max. stability pH 5.7	189
HYDROLASES				
Carboxyesterase	B. stearothermophilus	2 h, 65 °C	100% loss, 2 h, 75 °C	191
Carboxyesterase	B. stearothermophilus	10% loss: 2 h, 105 °C	Crude enzyme only moderately stable	192
Esterase	S. acidocaldarius	1 h, 100 °C	Stable 1 h, 80 °C	193
F 1, 6 bis Pase	F. thermophilum HB8	>30 min, 80 °C	Stable 1 h, 70 °C	197
Rest. endonuclease	S. solfataricus	30 min, 80 °C	$Sua1$	199
Endonuclease (AP)	Tx. thiopara	10 min, 70 °C	Stabilized further by BSA, $(NH_4)_2SO_4$	202
L-Asparaginase	T. aquaticus	40 min, 85 °C		204
D-Asparaginase	T. aquaticus	25 min, 85 °C	10% lost: 10 h, 70 °C	205
Arginase	"B. caldovelox"	105 min, 95 °C	Mn^{2+} and aspartic acid present (synergistic stabilization)	207
ATPase	S. acidocaldarius	st. 20 min, 89 °C	pH 6.6	211, 215
ATPase	S. acidocaldarius	20% loss: 10 min, 95 °C	solubilized enzyme	
ATPase	T. thermophilum HB8	100% loss: 10 min, 95 °C		217
α-Amylase	B. stearothermophilus	7 h, 85 °C; 1 h, 95 °C	Starch stabilizes	224
α-Amylase	B. 11-1S	2 h, 60 °C; 7.5 min, 70 °C	Ca^{2+}, EDTA no effect	225
Transglucosylase	S. solfataricus	>60 min, 90 °C		227, 228
Transglucosylase	Thermotoga sp.	20–30 min, 95 °C		
Transglucosylase	Tc. AN1	10–20 min, 95 °C		
Transglucosylase	Desulfurococcus sp.	4 min, 100 °C		

Enzyme	Organism	Conditions	Notes	Ref.
"Amylase"	*P. furiosus*	20% loss: 6 h, 100 °C	90% loss: 6 h, 120 °C	229
α-Glucosidase	*"B. caldovelox"*	1 h, 70 °C	Mn^{2+}, His, Cys stabilize further	230
Pullulanase	*Ta.* Tok6B1	17 min, 85 °C; 5 min, 90 °C		231, 232
Pullulanase	*T. aquaticus*	4.5 h, 95 °C		235
Pullulanase/amylase	*C. thermohydrosulfuricum*	st. 2 h; 65 °C; 70% loss: 10 min, 85 °C	Ca^{2+} stabilizes further	236, 237
Oligo-1,6-glucosidase	*B.* strain	10 min, 91 °C	Stable 15 h, 80 °C pH 6	239
Oligo-1,6-glucosidase	*Ta.* Tok 6B1	16.5 h, 70 °C; 20 min, 76 °C		240
CMCase	*"Cc. saccharolyticum"*	11 min, 90 °C	(hemi)cellulase components cloned into *E. coli*	250–259
Avicelase	*"Cc. saccharolyticum"*	20 min, 90 °C		250–259
β-Xylosidase	*"Cc. saccharolyticum"*	40 min, 70 °C		250–259
β-Glucosidase	*"Cc. saccharolyticum"*	70 min, 90 °C		260
β-Glucosidase	Cellulolytic organism	1 hr, 75 °C		261, 262
β-Glucosidase	*Thermus* sp.	5 d, 75 °C; 1 d, 80 °C		227, 228
β-Glucosidase	*Tc.* strain AN1	10 min, 95 °C		227, 228
β-Glucosidase	*Tc. celer*	~20 min, 105 °C		227, 228
β-Glucosidase	*Thermotoga* sp.	~20 min, 105 °C		227, 228
Xylanase	*B. stearothermophilus*	11 h, 74 °C, 30 min, 80 °C	Xylan stabilized further	265
Xylanase	*B.* strain	15 min, 80 °C		266
Xylanase	*Thermomonospora* strain	24 h, 65 °C		267
Xylanase	*Thermotoga* sp.	20 min, 105 °C	10 min, 115 °C; 1.3 min, 130 °C under certain conditions	227, 228, 268
β-Xylosidase	*"Cc. saccharolyticum"*	40 min, 70 °C	BSA, DTT stabilize	259
β-Galactosidase	*Thermus* 41A	40 h, 75 °C; 8 min, 90 °C		271
β-Galactosidase	*T. aquaticus*	~30 min, 90 °C		272
β-Galactosidase	*Ta.* strain	2.5 h, 75 °C		273
β-Galactosidase	*S. solfataricus*	23 d, 70 °C; 55 h, 85 °C	Partially purified	274
β-Galactosidase	*S. solfataricus*	30 d, 70 °C	Immobilized enzyme	275
β-Galactosidase	*S. solfataricus*	24 h, 75 °C; 3 h, 85 °C	Purified enzyme	276
β-Galactosidase	*D.* strain Tok12S1	8 min/70 min, 90 °C	2 component system?	277
β-Galactosidase	*Tc.* strain AN1	20 min, 90 °C		277
Protease	*B.* strain OK3	12 h, 75 °C; 40 min, 85 °C	Stable 7 days 75 °C	281, 282
Protease	*B.* strain EA.1	2 h, 85 °C		281, 282

(Continued page 12)

Table 1. Continued

Enzyme†	Organism*	Thermostability data**	Comments**	Ref.
Protease	B. stearothermophilus	40% loss: 1 h, 85 °C	Asporogenous mutant produced more enzyme	283
Protease	T. aquaticus T351	30 h, 80 °C; 1 h, 90 °C	4.8 min, 75 °C when treated with EDTA	284
Protease	T. aquaticus	4 h, 95 °C	La^{3+}-proteinase complex	285
Proteinase	T. aquaticus YT1	1 h, 80 °C	8 h, 80 °C in presence of Ca^{2+}	289, 290
Protease	"T. caldophilus"	40% loss: 20 min, 90 °C	Stable 1 h, 70 °C	292
Protease	Thermus ToK3	45 min, 90 °C; 5 min, 100 °C	Unstable low ionic strength – 80% loss: 1 h, 75 °C	293
Protease	Thermus Rt4.1A	13.5 h, 80 °C; 20 min, 90 °C		δ, 287
Protease	D. strain	80 min, 95 °C; 8 min, 105 °C	Mannitol, glycerol stabilize further	295
Protease	Tc. celer	40 min, 95 °C		227, 228
Protease	S. solfataricus	40 min, 95 °C		
Protease	S. acidocaldarius	st. 48 h, 80 °C	pH 4.5	297, 298
Aminopeptidase	T. aquaticus YT1	25% loss: 20 h, 80 °C		301
Aminopeptidase	S. solfataricus	90% loss: 15 min, 70 °C	Addition of Co^{2+} stabilizes: 24% loss: 15 min, 70 °C	302
Carboxypeptidase	T. aquaticus YT1	40% loss: 20 h, 80 °C		301
LYASES, ISOMERASES AND LIGASES				
Aldolase	T. aquaticus	2.5 h, 97 °C; 3.5 min, 105 °C		303
Citrate synthase	Tp. acidophilum	~10 min, 83 °C		305, 307
Citrate synthase	S. solfataricus	~10 min, 90 °C		
Fumarate hydratase	S. solfataricus	72% loss: 30 min, 90 °C	Stable 1 h, 85 °C	308
Threonine deaminase	T. strain X-1	15% loss: 1 h, 70 °C	30% loss: 1 h, 80 °C	309
Xylose isomerase	T. aquaticus HB8	4 d, 70 °C	Stable >1 month 70 °C with Mg^{2+}, Ca^{2+} or when immobilized	317

Enzyme	Organism	Stability	Conditions/Notes	
Tyr tRNA ligase	"B. caldotenax"	1 h, 70 °C	In PO_4^{2-} buffer, less stable in Tris	318
Arg tRNA synthetase	B. stearothermophilus	1 min, 74.5 °C	ATP and tRNA stabilize further	34
DNA ligase	T. thermophilus HB8	2 d, 65 °C		325
Glutamine synthetase	"B. caldolyticus"	5 min, 88–90 °C	In presence of Mn^{2+}, ATP and L-glutamate	77
Glutamine synthetase	B. stearothermophilus	90% loss: 3 h, 70 °C	Stable 5 h, 70 °C with Mg^{2+}/Mn^{2+}, glutamine and NH_4Cl	33

† Enzyme abbreviations

ADH:	alcohol dehydrogenase (1°: primary; 2°: secondary)
LDH:	lactate dehydrogenase
MDH:	malate dehydrogenase
isoPM DH:	threo-Ds-3-isopropylmalate dehydrogenase
GDH:	glucose dehydrogenase
G-3-P DH:	d-glyceraldehyde-3-phosphate dehydrogenase
tRNAMeTase:	tRNA methyltransferase
Asp amino Tase:	aspartate amino transferase
Phe aminoTase:	phenylalanine amino transferase
PFKinase:	phosphofructokinase
DNA-RNA poly.:	DNA dependent RNA polymerase
F-1,6-bisPase:	fructose-1,6-bisphosphatase
endonuclease (AP):	Endodeoxribonuclease (apurinic, apyrimidinic)
CMC ase:	carboxymethylcellulase

* Organism abbreviations

B.:	Bacillus	M.:	Methanothermus
C.:	Calderobacterium	Mc.:	Methanococcus
Cc.:	"Caldocellum"	P.:	Pyrococcus
Cl.:	Clostridium	S.:	Sulfolobus
D.:	Desulfurococcus	T.:	Thermus
F.:	Flavobacterium	Ta.:	Thermoanaerobium
		Tc.:	Thermococcus
		Th.:	Thermoanaerobacter
		Tp.:	Thermoplasma
		Tpr.:	Thermoproteus
		Tx.:	Thermothrix

** Values are half-lives (min, minutes; h, hours; d, days) at the given temperature unless otherwise indicated, i.e. st, stable (no activity loss) or % loss

δ Peek K et al. unpublished data

NAD(P)H needed for such reactors, the life-time of which would be extended with stable enzymes [see e.g. 95]. With regard to biosensors, the possibility of oxidation of NADH (resulting from dehydrogenase activity) with concomitant hydrogen peroxide production is an important reaction due to the comparative ease of detection of hydrogen peroxide. An $NADH:H_2O_2$ oxidoreductase has been recently demonstrated in a *Thermus* isolate [96] which has activity over a broad pH range and at high temperatures (up to 95 °C), although thermostability data were not given. Another enzyme with potential application for regeneration of $NAD(P)^+$ and NAD(P)H is the viologen-dependent pyridine nucleotide oxidoreductase isolated from an extremely thermophilic strain of *Bacillus stearothermophilus* by Nagata et al. [97].

The alcohol dehydrogenase from *Thermoanaerobium brockii* has been shown to reduce aliphatic acyclic ketones asymmetrically, optimal optical purity being obtained (at the expense of reaction rate) at temperatures in the ambient range [98]. Further, this enzyme has been used to obtain optically pure cyclic ethers which are constituents of civet used as a fixative in the perfume industry [99].

An archaebacterial NAD^+-dependent (and possibly zinc requiring) alcohol dehydrogenase from *Sulfolobus solfatarius* with tolerance to high temperature and organic solvents is also known [100] and has been purified to homogeneity with 50% recovery using glycerol to prevent aggregation. The initial rate of reaction of this enzyme was reported to increase over the range 45–90 °C, but thermostability was not particularly high considering the growth temperature of the organism. This may be a result of suboptimal stabilization of the enzyme during purification procedures. The $NADP^+$-dependent secondary alcohol dehydrogenase from the eubacterium *Thermoanaerobacter ethanolicus* has also been studied [101, 102]. Whereas the secondary alcohol dehydrogenase could act on ethanol, albeit at a rate 15 times slower than that on the preferred substrate propan-2-ol, the primary alcohol dehydrogenase was specific to primary alcohols, oxidising heptanol at a similar rate to ethanol. The specificity of the *T. ethanolicus* secondary alcohol dehydrogenase is similar to that described for the corresponding enzyme from *Thermoanaerobium brockii* [103]. This enzyme was shown to prefer secondary alcohols (pH optimum 7.8–9.0) but it was also active on primary alcohols, ketones and acetaldehyde (pH optimum 7.8) [104].

Lactate dehydrogenases are perhaps the best studied of the oxidoreductases from extremely thermophilic organisms, including those of *Thermus* isolates [105, 106], *Bacillus stearothermophilus* [107–112], *Clostridium thermohydrosulfuricum* [113] and *Thermotoga maritima* [114]. The studies on *B. stearothermophilus* lactate dehydrogenase include work on the gene sequence [115] and site directed mutants [116]. The lactate dehydrogenase of an isolate of the genus *Thermus* was found to be virtually inactive above approximately pH 5.5 in the absence of the effector fructose-1,6-bisphosphate. In the presence of the effector the enzyme was active at neutral pH [105] and was also stabilized. The *Thermotoga maritima* lactate dehydrogenase has been reported as being a tetrameric enzyme of 144 kDa with identical subunits and high homology with the *Thermus* counterpart [114]. The enzyme had a broad pH optimum of 6.0–8.0, but was most stable at pH 6.0. Stability was enhanced by NAD^+ and fructose-1,6-bisphosphate, but decay at

90 °C in the presence of these components was apparently not first order. The Km values of the enzyme at 55 °C for pyruvate and lactate were 3.7 mM and 410 mM respectively, these being lowered to 0.06 mM and 25 mM respectively by the effector. Malate dehydrogenases have been studied in both eubacterial and archaebacterial systems, including those from *"Thermus flavus"* [73, 117–120] and *Thermus aquaticus* [121]. The latter enzyme is the most stable malate dehydrogenase yet reported, the data of Smith and Sundaram [121] corroborating the earlier findings of Pask-Hughes and Williams [122] on various *Thermus* isolates from Iceland. The corresponding enzyme from *"Bacillus caldotenax"* (a. *B. stearothermophilus* variant) is less stable [121] and is tetrameric, as opposed to the *Thermus* enzymes, which, in common with most malate dehydrogenases [123] are dimeric. With regard to archaebacterial sources, the malate dehydrogenase from *Thermoplasma acidophilum* [124] and from *Sulfolobus acidocaldarius* [125] are less intrinsically stable than might have been expected.

The malate dehydrogenases of the archaebacteria mentioned above are unusual in that they are also tetrameric and use both NADH and NADPH as coenzyme in oxaloacetate reduction (although Görisch et al. [126] have shown stereoselectivity to be the same as for the corresponding eubacterial enzymes). In *Thermoplasma acidophilum* the NADPH-dependent activity was only 10% that of the NADH-dependent activity [124]. This may be pH-dependent, since with *Sulfolobus acidocaldarius* malate dehydrogenase NADH was the preferred cofactor at near neutral pH (by a factor of 5) yet at more acid pH values NADPH gave the higher reaction rates [125]. A third archaebacterial malate dehydrogenase which has been studied is that from *Methanothermus fervidus* and is apparently a dimeric enzyme with similar Km and V_{max} values with both NAD^+ and $NADP^+$ [127]. The Arrhenius plots obtained for this enzyme showed discontinuities at about 52 °C and 70 °C, the latter being consistent with the low stabilities of the other archaebacterial malate dehydrogenases and the former suggested by Honka et al. [127] as reflecting the change in catalytic properties of the enzyme to the physiological temperature range. Earlier studies on *M. fervidus* reported on the thermostability of the malate dehydrogenase and its stabilization by the potassium salt of cyclic 2,3-diphosphoglycerate, the unusual trianionic compound found in the organism [128].

It may be that dual specificity for NAD^+ and $NADP^+$ is a general archaebacterial feature, since it is also seen for other dehydrogenases in these organisms, including D-glucose dehydrogenase [129], glyceraldehyde-3-phosphate dehydrogenase [130] (see below) and isocitrate dehydrogenase [131]. This last enzyme, isolated from *Sulfolobus acidocaldarius*, was shown to have a Km for NAD^+ approximately two orders of magnitude greater than that for $NADP^+$, the V_{max} for both nucleotide dependent activities being similar [131].

Malic enzyme has been isolated from the archaebacterium *Sulfolobus solfataricus* and has been shown to be essentially $NADP^+$ dependent and is markedly thermostable [132]. It was activated by NH_4^+, but there was no absolute requirement for the ion and its presence decreased the affinity of the enzyme for both malate and $NADP^+$ [132]. The high resistance of the enzyme to organic solvents and general protein denaturants has been studied in some detail [133]. Typically the enzyme retained full activity after 24 h at 25 °C in 4 M urea and 50% dimethyl-formamide, but was less stable in other chaotropic agents or at higher temperatures. A stimulatory effect on activity was observed with certain solvents, this being ascribed to an alteration in the V_{max} of the enzyme. The correlation between both activity and stability of the enzyme with the hydrophobicity of the solvent mixtures was investigated [133]. The potential use of both this enzyme and the alcohol dehydrogenase from *S. solfataricus* for regeneration of NAD(P)H in continuous flow membrane reactors has been investigated [134].

Isocitrate dehydrogenases have been studied in extreme thermophiles mainly from the point of view of elucidation of mechanisms of thermostability. This has included the studies of Hibino et al. [32] on the enzyme from a *Bacillus stearothermophilus* strain grown at 65 °C. In the presence of substrate the enzyme was markedly stabilized as judged by retention of tertiary conformation, reflected by the dramatically improved resistance to denaturation by urea. The thermo-stability of the partially purified isocitrate dehydrogenase from *Thermus aquaticus* was shown to be dependent upon the buffer used and on the addition of isocitrate [135]. This enzyme has also been purified from *Thermus thermophilus* HB8 [136]. The enzyme had a dimeric structure with identical 57.5 kDa subunits as judged by SDS-polyacrylamide gel electrophoresis, whilst gel filtration of the native enzyme on two different matrices gave values of 95 kDa and 120 kDa. The enzyme had a pH optimum of 7.8, required a divalent metal ion for activity ($Mn^{2+} > Mg^{2+}$) and was $NADP^+$ dependent, although NAD^+ was shown to replace $NADP^+$ with low efficiency.

The *threo*-Ds-3-isopropylmalate dehydrogenase from *T. thermophilus* HB8 has also been studied, again from a view to elucidate mechanisms of stabilization. Purified after first cloning into *E. coli* [137], the enzyme was shown to be dimeric with an optimum pH of 7.2 and a dependence upon metal ions (Mg^{2+} or Mn^{2+}) for activity. The enzyme was markedly stabilized by high concentrations (2 M) of KCl, but the mechanism for this is unclear.

Vali et al. [138] found the alanine dehydrogenase from *Thermus thermophilus* to be similar to a mesophilic counterpart from *Bacillus subtilis* in terms of catalytic parameters, but of greater stability.

Glucose dehydrogenase is of interest commercially for a single step glucose assay. The enzyme isolated from *Sulfolobus solfataricus* [139] was only moderately thermostable but had a desirably high degree of specificity for glucose when using NAD^+ as coenzyme (use of $NADP^+$ resulting in relaxation of specificity and reduced oxidation rates), and a relatively good stability in solvents.

D-Glyceraldehyde-3-phosphate dehydrogenases have been isolated from several extremely thermophilic eubacteria and archaebacteria including *Bacillus stearo-thermophilus* and *Thermus* isolates [140, 141], *Methanothermus fervidus* [142] and *Thermoproteus tenax* [143]. The *Bacillus stearothermophilus* and *Thermus aquaticus* NAD^+-dependent enzymes have been extensively studied, the holo and apo forms of both enzymes having been purified [141], amino acid sequences determined [144, 145] and their mechanisms of thermostabilization investigated [146]. The corresponding enzyme from *Thermotoga maritima*, which has 60% amino acid sequence homology to the *Bacillus* and *Thermus* enzymes [70], is an NAD^+-

dependent enzyme with four identical subunits [71]. The pH optimum for the oxidative reaction was temperature dependent, being in the range 8.5–9.5 at 40 °C and shifting to lower values at higher temperatures, e.g. 6.0–8.0 at 60 °C. The low temperatures were used to facilitate kinetic studies with unstable substrates and cofactor. Differential scanning calorimetry of the enzyme showed a melting temperature (Tm) for the protein of 109 °C in the presence of NAD^+ (partial removal of which lowered the Tm), indicating greater stability for this dehydrogenase than any other investigated oligomeric enzyme. In contrast to the eubacterial enzymes, the glyceraldehyde-3-phosphate dehydrogenase from the archaebacterium *Methanothermus fervidus* uses both NAD^+ and $NADP^+$ as cofactor, although again the Km for the former was markedly higher than for the latter [142] and thermostability was improved only by $NADP^+$. Subsequent studies on this enzyme (from the point of view of determining mechanisms of thermoadaptation) showed that the predominant trianionic compound cyclic 2,3-diphosphoglycerate in the organism stabilized the enzyme such that the half-life at 90 °C was extended by 140-fold [128].

The archaebacterium *Thermoproteus tenax* is unique in that it has been reported [143] as producing two distinct glyceraldehyde-3-phosphate dehydrogenases, one specific for NAD^+ and the other for $NADP^+$, rather than possessing a single enzyme with dual cofactor specificity. The NAD^+ specific enzyme was competitively inhibited by $NADP^+$ whereas the reverse was not observed. These two enzymes are the most stable of their type reported to date. The *Pyrococcus woesei* glyceraldehyde-3-phosphate dehydrogenase was shown to be similar to the *M. fervidus* enzyme in utilizing both NAD^+ and $NADP^+$ [147]. Although this enzyme was extremely stable at 100 °C, further stabilization was attained by the addition of salts particularly potassium citrate. Structurally the enzyme proved to be unusual in its high phenylalanine content and low proportion of aspartic acid, methionine and cysteine [147].

Glutamate dehydrogenase has been found in high levels (of the order of 5% of total protein) in the archaebacteria *Sulfolobus solfataricus* (Consalvi, pers. comm.) and *Thermococcus* strain AN1 (Hudson R C and Daniel R M: unpublished observations). For the latter enzyme the Km values for NAD^+, $NADP^+$ and glutamate at 80 °C were found to be 0.05 mM, > 100 mM and 11 mM respectively. At 60 °C, however, substantially lower Km values were found for both NAD^+ (0.008 mM) and glutamate (2 mM). The enzyme was stimulated several fold by Mg^{2+}, Ca^{2+} and Mn^{2+} (all at 50 mM).

The NADH dehydrogenase isolated from the archaebacterium *Sulfolobus acidocaldarius* is specific for NADH and reduces certain quinones, ferricyanide and 2,6-dichlorophenolindophenol [148]. The actual pH optimum of the enzyme was dependent upon the electron acceptor used, being considerably lower for ferricyanide (4.5) than for 2,6-dichlorophenolindophenol. The membrane enzyme from "*Bacillus caldotenax*" had similar properties when solubilized and partially purified, activity being stimulated by membrane lipids and non-ionic detergent [149]. The "*Bacillus caldovelox*" enzyme was more stable in the membrane bound form than when solubilized and the addition of membrane lipids to the soluble enzyme preparation resulted in restoration of thermostability to levels comparable to the membrane bound form. The corresponding enzyme from a *Thermus aquaticus* isolate [83, 150] had a pH optimum of 3.6 with ferricyanide and exhibited good stability.

Hydrogenases, useful in redox cofactor regeneration, have been found in several extreme thermophiles. Pinkwart et al. [151] isolated the enzyme from *Bacillus schlegelii* grown at 70 °C, reporting it to be oxygen sensitive but not giving any thermostability data. Hydrogenase has also been studied in *Calderobacterium hydrogenophilum* [152] and shown to have a half-life of 4.5 min at 100 °C [153]. Shah and Clark [154] have partially purified two hydrogenases from *Methanococcus jannaschii*. The F_{420}-nonreactive hydrogenase comprised two subunits whereas the F_{420}-reactive enzyme comprised three subunits. The pH optimum for the latter enzyme was substrate dependent and could not be determined for methyl viologen due to enzyme instability at high pH values over which activity against the substrate was still increasing. The F_{420}-nonreactive hydrogenase exhibited methyl viologen-reducing activity over a broad pH range with an optimum at about pH 9. The F_{420}-nonreactive enzyme was more thermostable than the F_{420}-reactive enzyme and has been shown (Clark, pers. comm.) to be further stabilised by glycerol. Activity loss was not first-order (as was the case with the *C. hydrogenophilum* enzyme), with some activity remaining after 5 min at 115 °C.

The hydrogenase from *Pyrococcus furiosus* has also been studied [155] and found to be a heterotrimeric enzyme ($\sim$185 kDa) which preferentially catalysed H_2 production, especially over the growth temperature range of the organism. The enzyme was extremely stable under anaerobic conditions, but was destabilized by dithionite and, more severely, under aerobic conditions.

4.2 Transferases

4.2.1 Nucleotidyl Transferases

Of the thermostable transferases, the nucleotidyl transferases have received the most attention in recent years. This is due to the interest in DNA polymerases which mediate the polymerisation chain reaction [156], i.e. the amplification of sequences of DNA. This reaction forms the basis of a number of commercial applications in the fields of medicine and forensic science, as well as in research in molecular biology (see Oste [157]).

The use of a thermostable DNA polymerase allows both simplification of the assay and improved specificity. The former is the result of a thermostable enzyme surviving the heating cycle for DNA strand separation and therefore obviating the addition of fresh enzyme (as has been necessary with mesophilic DNA polymerase, i.e. Klenow fragment from *E. coli*), and simplifying automation. Improved specificity is a consequence of being able to run the polymerase chain reaction at higher temperatures, thus ensuring minimum non-specific binding of the primer. These observations have been made by Saiki et al. [158] using DNA polymerase from *Thermus aquaticus*, comparing its performance in the amplification reaction with that given using Klenow fragment.

Single DNA polymerases were isolated from *Thermus ruber, T. "flavus"* and *T. aquaticus* strain TYI by Kaledin et al. [159–161]. The *T. ruber* enzyme was purified to homogeneity and did not exhibit exonuclease activity. It showed a preference for Mg^{2+}, but little data on stability were given. The enzymes from both *"T. flavus"* [160] and *T. aquaticus* [159] were also devoid of exonuclease activity. The former enzyme was Mg^{2+} dependent with a pH optimum of about 10, whereas the pH optimum of the latter DNA polymerase was 8.3.

Rüttimann et al. [162] reported the presence of three DNA polymerases, designated A, B and C, in *Thermus thermophilus* strain HB8. All were free of exonuclease activity but differed in chromatographic behaviour, thermostability (reflected by temperature dependence in in vitro assays), specificity and divalent metal ion response. The thermal decay curves for the A and C enzymes were biphasic. The presence of three DNA polymerase isozymes in this organism, as opposed to the single enzyme in those organisms mentioned above and in the *Thermus aquaticus* strain studied by Chien et al. [163], was thought by Rüttimann and colleagues to be a consequence of improved chromatographic separation procedures. *"T. flavus"* cells were in fact found to contain three DNA polymerases using phosphocellulose columns [164]. Work in this laboratory on *T. aquaticus* T351 also indicates the presence of more than one DNA polymerase (Coolbear T, unpublished observations). The *Thermotoga* strain FjSS3-B.1 may also possess more than one DNA polymerase, but purification of the enzyme from this organism proved difficult, with the affinity matrices used routinely for other eubacterial and archaebacterial DNA polymerases proving unsuccessful for the enzyme [164]. The DNA polymerase from *Thermotoga* strain FjSS3-B.1 was a monomeric, 85 kDa enzyme with a pH optimum of 7.5–8.0 and a divalent metal ion requirement. The enzyme was only moderately thermostable in the absence of substrate, but was stabilized by mannitol and Triton X-100, the effect of these agents being additive [164].

Archaebacterial DNA polymerases have been isolated from a number of organisms. The enzyme from *Methanobacterium thermoautotrophicum* [165] is a single polypeptide (72 kDa) in the purified form with Mg^{2+} dependence for activity and stimulated by KCl (albeit at lower concentrations than the intracellular levels reported by Sprott and Jarell [166]). Thermostability data were not given, although the enzyme gave highest activity at 65 °C in 30 min assays with about 15% of this activity at 80 °C.

Sulfolobus species also appear to contain more than one DNA polymerase. The enzyme reported by Rossi et al. [167] from *S. solfataricus* was the major component of two chromatographically separable DNA polymerase activities, and showed a preference for Mg^{2+} for activity. Loss of activity of the enzyme above 75 °C in short assay periods was probably due to DNA template melting rather than thermostability problems. Three discreet DNA polymerases, designated A, B and C, were at least partially purified from *S. acidocaldarius* [168], an organism also used as a source of DNA polymerase by Klimczak et al. [169]. The A and B enzymes were unaffected by aphidicolin and were relatively thermostable.

DNA dependent RNA polymerases have been studied in the extremely thermophilic methanogens *Methanobacterium thermoautotrophicum* strain W, *Methano-*

thermus fervidus and *Methanococcus thermolithotrophicus*, and were compared to the enzymes from other lower temperature methanogens [170]. The three enzymes were all multicomponent and were dependent on divalent metal ions for activity, Mg^{2+} being more effective than Mn^{2+}. The temperatures required for maximum rates of transcription in vitro were well below the growth temperatures of the organisms. The *Methanothermus fervidus* enzyme, for example, was inactive at 85 °C yet the organism was grown at 83 °C.

The DNA-dependent RNA polymerases from the sulphur-dependent archaebacteria *Sulfolobus acidocaldarius*, *Thermoproteus tenax* and *Desulfurococcus mucosus* have been compared [171]. The enzymes comprised 9–10 subunits and were resistant to rifampicin and streptolydigen, as was the case with enzyme from *Methanobacterium thermoautotrophicum* [172]. The *T. tenax* enzyme was the most stable. Some early work on DNA dependent RNA polymerases was also undertaken on *Thermus* isolates, the *Thermus aquaticus* enzyme being studied by Air and Harris [173] and Fabry et al. [174]. In 10 min assays greatest activity was obtained at 65 °C, but no thermostability data were given. Using polymin P precipitation and FPLC to purify the enzyme, Wnendt et al. [175] have shown that the DNA-directed RNA polymerase from *Thermus thermophilus* HB8 is also a multi-component enzyme with β, β', α and σ subunits. Thermine was reported to activate the enzyme eight-fold, and two nucleotide binding sites with strong positive cooperativity were identified.

4.2.2 Other Transferases

tRNA methyltransferases have been studied in *Thermus* strains [176–178]. The enzyme acting on adenine from "*T. flavus*" was shown to be Mg^{2+} dependent, with stimulation of activity given by low concentrations of Na^+ and NH_4^+, and exhibited a pH optimum of near neutrality [176]. The tRNA-(guanosine-2'-)-methyltransferase from *T. thermophilus* HB27 [177, 178] has similar thermostability to the *T. flavus* enzyme, as has the tRNA-(adenine-1-)-methyltransferase from a second strain of this organism, *T. thermophilus* HB8 [179].

One of two novel transferases reported recently is the tetrahydromethanopterin methyltransferase involved in methanogenesis in *Methanobacterium thermoautotrophicum* [180]. No thermostability data on the oxygen sensitive enzyme were presented. The second enzyme is the propylamine transferase studied in the archaebacterium *Sulfolobus solfataricus* [181]. This enzyme (optimum pH 7.5) appears to be solely responsible for transfer of the aminopropyl residue from *S*-adenosyl (5')-3-methylthiopropylamine to various acceptors.

Aspartate aminotransferase from *S. solfataricus* has been shown to be a dimeric enzyme with unique coenzyme binding properties [182]. It is extremely thermostable with a melting temperature of 109 °C [183] and an "optimum" temperature for activity of 100 °C. The enzyme has been cloned and studied from an evolutionary viewpoint [184].

L-Phenylalanine aminotransferases (classified as tyrosine aminotransferase, E.C. 2.6.1.5 which also acts on phenylalanine) are enzymes useful in L-phenylalanine synthesis, as mentioned above, and have been studied in various thermophilic bacteria by Schutten et al. [185]. These workers looked at several *Bacillus* isolates and *Thermus aquaticus* strain YT-1, reporting that the most stable L-phenylalanine aminotransferases were to be found in two *Bacillus stearothermophilus* variants selected for by using L-phenylalanine as sole carbon source at growth temperatures of 70 °C and 75 °C.

Early work on phosphofructokinase from extreme thermophiles was undertaken from a structural and mechanistic viewpoint. Hengartner and Harris [186] obtained the enzyme from both *Thermus aquaticus* and a *Bacillus stearothermophilus* strain grown at 60 °C. The enzyme from the former organism was shown to be slightly more stable than the phosphofructokinase from *Flavobacterium thermophilum* HB8 [187]. Xu et al. [188] have studied the allosteric nature of phosphofructokinase from *Thermus thermophilus* HB8. In the presence of phosphoenolpyruvate the enzyme existed as an inactive dimer, being reverted to the active tetramer by the addition of substrate or cofactors. Pyruvate kinase has also been partially purified from a *Thermus thermophilus* isolate and studied from the viewpoint of glycolytic regulation [189]. Stability was maximal at pH 5.7, a value similar to that for optimum activity.

ATP sulphurylase has been purified from the sulphate reducing archaebacterium *Archaeoglobus fulgidus* [190]. The enzyme was tentatively assigned an $\alpha_2\beta$ structure (150 kDa) which tended to aggregate into active trimers, probably through hydrophobic interactions. The enzyme had a pI of 4.3 and exhibited optimum activity at pH 8.0. Activity of the crude enzyme was "optimal" at 90 °C (no time factor was given) and the purified enzyme was not studied. No thermostability data were given.

4.3 Hydrolases

4.3.1 General Hydrolases

Esterases are being increasingly recognised as useful for stereospecific manipulation of esters, but little is known about these enzymes in extreme thermophiles. Two esterases from *B. stearothermophilus* variants grown at 65 °C have been reported [191, 192]. The first [191] was a monomeric enzyme of 42–47 kDa with a pH optimum of 7.0 and a susceptibility to inhibition by serine reactive agents. Inhibition was also observed with sulphydryl reagents, but this may have been a result of destabilization. The Km of the enzyme decreased as chain length of fatty acyl substrates increased, but V_{max} was highest with *n*-caproate. The initial rate of reaction with *p*-nitrophenyl *n*-caproate was found to be maximum at 65 °C. The other *Bacillus* carboxyesterase [192] was significantly more stable. The monomeric enzyme, of 38–45 kDa and pH optimum 9.0–9.5, was less sensitive to serine reactive agents and retained the majority of its activity after 2 h of incubation at 105 °C. This contrasted markedly with the relatively moderate thermostability of the enzyme in the crude form. A number of possible explanations were given [192] but none was confirmed.

One archaebacterial esterase has been purified from *Sulfolobus acidocaldarius* [193]. This enzyme was assessed as being a tetrameric enzyme with 32 kDa subunits, a pH optimum of 7.5–8.5 and sensitivity to serine reagents. The V_{max} of the enzyme was highest with C_5 acyl substrates, but the Km of the enzyme for this substrate was higher than against other longer or shorter chain molecules tested.

Alkaline phosphatases have been isolated from both *Thermus aquaticus* [194, 195], *Thermus* strain Rt4.1A (Hartog A, Daniel R M: unpublished observations) and *T. thermophilus* [196]. For the latter enzyme a pH optimum of 10.5 was observed and maximum activity was given at 70 °C in 10 min assays. The alkaline phosphatase reported by Yeh and Trela [194] had a pH optimum of 9.2 and in 10 min assays maximum activity was given at about 75 °C. The work of Smile et al. [195] showed that *T. aquaticus* alkaline phosphatase possessed phosphodie-sterase as well as monoesterase activity. At the optimum pH (7.2), diesterase activity was maximal at 80–85 °C in 10 min assays, a higher temperature (and lower pH optimum) than found for the monoesterase activity. No further data pertaining to the effect of temperature on activity or thermostability were given in either study. *Flavobacterium thermophilum* HB8 has been shown to contain a D-fructose-1,6-bisphosphatase which was studied from the perspective of regulation of glycolysis [197].

A limited number of restriction endonucleases have been purified from extreme thermophiles. One of the earliest studies on these enzymes was on the *Taq* I endonuclease from *Thermus aquaticus*, but no thermostability studies were made [198]. *Bacillus* derived type II endonucleases include *Bcl* I from *B. "caldolyticus"* [199] and *Bst* 1503 from a moderately thermophilic strain of *B. stearothermophilus* [200]. An archaebacterial type II endonuclease *Sua* I from *Sulfolobus solfataricus* has also been described [201].

Kaboev et al. [202] reported on an enzyme (specific to apurinic and apyrimidinic sites in DNA) from *Thermothrix thiopara* growing at 75 °C in spring water or cultured at 70 °C in the laboratory. The enzyme did not appear to be metal ion dependent, being insensitive to EDTA and its properties differed from those of the enzyme from *B. stearothermophilus* [203]. Despite the proposed importance of the enzymes in repair of depurinated DNA (whether spontaneously or by the action of DNA glycosylases) and the consequence of higher temperatures on the rate of depurination, few other thermophilic endonucleases of this type have been studied.

There are few reports of hydrolases from extreme thermophiles acting on non-peptide bond C-N groups. D-Asparaginase from *Thermus aquaticus* has been studied by Guy and Daniel [204] who found that the enzyme had a histidine residue at the active site, contained six disulphide bridges per molecule and was relatively thermostable. Curran et al. [205] presented the first report of an L-asparaginase from an extreme thermophile. These workers found that the enzyme from *Thermus aquaticus* strain T351 was highly specific to L-asparagine, but had a high Km (8.6 mM) and was therefore not considered to be useful as a potential antineoplastic agent.

In preliminary screening studies [206] arginases have been observed in a number of extreme thermophiles, namely representatives from the three species of *Thermus*, from *Bacillus stearothermophilus* strains and two archaebacterial *Sulfolobus* isolates, *S. acidocaldarius* and *S. solfataricus*. In the same work arginine deiminase and arginine decarboxylase activities were not detected in these organisms, but were found to be present in other archaebacteria and *Clostridium thermo-hydrosulfuricum*. More extensive work on the arginase from the *Bacillus stearo-thermophilus* variant "*B. caldovelox*" showed the enzyme to be a highly thermo-stable hexamer with divalent cation requirement for activity [207].

Inorganic pyrophosphatases have been purified from both a *Bacillus stearo-thermophilus* isolate grown at 65 °C [208] and "*Thermus flavus*" 70 K [209]. The *Thermus* enzyme was capable of hydrolysing ATP as well as pyrophosphate and was apparently stable over a period of 2 h at 70 °C although activity was lower than at 60 °C. When pre-incubated at 60 °C and subsequently assayed at 40 °C enzyme activity was markedly enhanced [209]. The corresponding enzyme from *Thermus aquaticus* was found to depend on Mg^{2+} for activity which was maximal at 80 °C in 5 min assays at pH 8.3, but no information on thermostability was given [210]. Wakagi and Oshima [211] reported the presence of inorganic pyrophosphatases in *Sulfolobus acidocaldarius*, but this study concentrated on the ATPases of the archaebacterium and no detailed data on the enzymes were given, save that they had similar pH optima and stabilities to the ATPases (see below).

It has been established that archaebacterial membrane ATPases are genetically related to eukaryotic vacuolar H^+ ATPases (rather than the membrane F_0F_1AT-Pases) and together are termed V-type ATPases [212, 213]. The two ATPases identified in *Sulfolobus acidocaldarius* strain 7 were both resistant to inhibition by dichlorohexylcarbodiimide and azide [211]. One of the enzymes had a pH optimum of 2.5 and was stable in the absence of exogenous metal ions. The other enzyme, similar in some respects to the *Thermoplasma* ATPases [214], had optimal activity at pH 6.5 and was dependent for stability on divalent metal ions (Mg^{2+} or Mn^{2+}). Subsequent work on ATPase solubilised from the membranes of *S. acidocaldarius* strain 7 showed it to comprise three different subunits [215]. It was activated by bisulphite and gave maximum activity at 85 °C in 10 min assays at pH 5.5. The corresponding enzyme in a different strain of *S. acidocaldarius* (DSM 639) was reported by Lübben and Schäfer [216] to be composed of two major subunits. Inhibitor responses were similar to the enzyme from *S. acidocaldarius* strain 7. Two pH optima of 5.5 and 8.0 were, however, exhibited by the enzyme in the absence of bisulphite, a greater level of activity with a single pH optimum of 6.2 being given upon the inclusion of the salt. This was interpreted as a preferential stimulation of activity at this pH. The enzyme required Mg^{2+} or Mn^{2+} for maximum activity, but no data on stability were given. A non-specific inorganic pyrophosphatase with a pH optimum of 2.9 was also identified in *S. acidocaldarius* strain DSM 639 and this possibly accounted for the acid ATPase in strain 7 [216].

An ATPase has recently been described in *Thermus thermophilus* strain HB8 [217]. On the basis of azide insensitivity, nitrate sensitivity, a large α-subunit and N-terminal amino acid sequences of the two major subunits, the enzyme is probably a V-type ATPase. The pH optimum for activity was 7.5 and maximum activity was obtained at 85 °C in 5 min assays. Mn^{2+} stimulated activity, and sulphite increased activity 30-fold at the optimum pH and 50-fold at pH 6.5.

4.3.2 Glycoside Hydrolases

Amylolytic enzymes are widely used in the starch industries, and have been reviewed extensively in recent years [218–223].

α-Amylases are some of the most stable enzymes from mesophiles. The industrial use of *B. licheniformis* α-amylase is based on its stability for short periods at 110 °C in the presence of starch and calcium and its prolonged resistance to denaturation at 95 °C. Grueninger et al. [224] reported on the α-amylases from *B. stearothermophilus* strains isolated from heat-treated sewage sludge, these organisms growing optimally at up to, or in excess of, 70 °C. The presence of starch again markedly stabilized the enzyme. Although the α-amylases from extreme thermophiles have not necessarily matched this level of stability, other properties of some α-amylases from certain organisms are of interest. For instance, the acidophilic *Bacillus* strain 11-1S reported to grow optimally at 65 °C in the pH range 3.0–4.0 possesses an α-amylase with a pH optimum of 2.0 [225].

Plant et al. [226] screened five strains of *Clostridium thermohydrosulfuricum*, two *Fervidobacterium* sp., two *Thermoanaerobium* sp., one *Thermoanaerobacter* sp. and one *Dictyoglomus* sp. for α-amylases. None of the enzymes was characterised however, and no data on thermostability are available. The organisms were all anaerobes with growth optima above 65 °C in the pH range 6.5–7.5. Amylolytic activity has also been observed in a survey of 14 *Desulfurococcus* strains (including *Thermococcus thermophilum*), *Sulfolobus solfataricus*, *S. acidocaldarius* and in four *Thermotoga* strains [227, 228]. In *Thermococcus* strain AN1 the amylolytic activity is due to a combination of α-glucosidase, pullulanase and transglucosylase activities. Recently the archaebacterium *Pyrococcus furiosus* has been reported as having amylolytic activity with a pH optimum of 5.0, a temperature "optimum" of 100 °C (no time factor being given) and extremely high thermostability [229]. The enzyme was reported to degrade starch and amylase in a random fashion, whilst pullulan and maltose were not hydrolysed.

The α-glucosidases (maltases) are useful in the production of glucose from starch in the presence of α-amylases since they hydrolyse 1,4-α-glucosidic linkages (and, more slowly, 1,6 bonds) in the short chain oligosaccharides which are products of α-amylase action. A recent example of this enzyme for an extreme thermophile is that from a strain of *B. stearothermophilus "caldovelox"* [230]. The enzyme, maximally active at pH 5.5–6.0 was only moderately thermostable. Activity was enhanced over 4-fold by the inclusion of manganous ions or by chelating agents. This latter phenomenon remained unexplained, but may be parallelled by the presumed effect of chelators on the equilibrium of toxic and non-toxic divalent metal ion configurations postulated for manganese interaction with arginase [206].

Pullulanase (EC 3.2.1.41) is the enzyme which hydrolyses pullulan, a polymer of 1,6-linked maltotriose units, to maltotriose. This distinguishes it from other enzymes that can act on pullulan to give isopanose, panose or glucose, i.e. isopullulanase (EC 3.2.1.57), α-amylase (EC 3.2.1.1) and glucoamylase (EC 3.2.1.3) respectively. Plant et al. [231, 232] studied the pullulanase from *Thermoanaerobium* strain Tok 6-B1 and found it to hydrolyse not only the α1,6 bonds in pullulan but also the branch point α1,6 bonds of amylopectin and glycogen. Further, the enzyme hydrolysed α1,4 bonds of long chain malto-oligosaccharides in a random, endo fashion, as well as short chain malto-oligosaccharides (up to 7 residues), giving maltose as the minimum product (not glucose). The enzyme had similar stability to the pullulanases from *Clostridium thermohydrosulfuricum* [233] and the *Bacillus stearothermophilus* enzyme [234], the former enzyme being stabilized by starch. The pullulanase described in *Thermus aquaticus* YT-1 [235] was significantly more stable than the *Thermoanaerobium* enzyme. Melasniemi [236] reported that the α-amylase and pullulanase activities in *Clostridium thermohydrosulfuricum* were properties of a single protein. The enzyme was subsequently purified in two forms [237] with similar properties apart from different mobilities on gel electrophoresis (albeit in diffuse bands). A review on the novel pullulanases cleaving α1,4 linkages in starch has recently been published [238].

An *exo*-oligo-1,6-glucosidase, cleaving single glucose residues from the non-reducing terminal α1,6 linkages of panose, small isomalto-oligosaccharides and α-limit dextrins, has been isolated from an extremely thermophilic *Bacillus* strain grown at 80 °C [239]. The enzyme was apparently most stable at pH 6, whereas the optimum assay pH was approx. 5.0.

An oligo α1,6-glucosidase has also been observed in the extremely thermophilic anaerobe *Thermoanaerobium* strain Tok 6-B1 [240]. The enzyme was active against α1,6 bonds in isomaltose, isomaltotriose and panose, but not against malto-oligosaccharides, pullulan or amylose, and is therefore similar in its α1,6 specificity to the *Bacillus thermoglucosidus* enzyme [241, 242]. The *Thermoanaerobium* enzyme had optimal activity at pH 5.6–7.0 and was considerably more stable than the *Bacillus* enzyme.

Cellulosic biomass is a major waste product, and yet is a renewable source of chemical feedstock. The cellulases acting on this material can be considered as having two main uses, either a direct use in food processing (for humans and other animals), pharmaceutical processes and sewage treatment, or in the production of sugars as energy sources for fermentative production of value-added products [243]. Ethanol production via glucose (both by cellulolytic organisms and enzyme systems) and selective degradation of wood pulp cellulose and hemicellulose to give high strength papers are typical examples of cellulase applications.

The cellulases from thermophilic organisms have been reviewed in detail recently by Margaritis and Merchant [244] (see also [245, 246]) whose article contains information on organisms growing at 45 °C or above, this criterion covering the cellulases produced constitutively by *Clostridium thermocellum*. These organisms generally have growth optima of 60 °C and thus lie outside the general scope of this review.

C. thermocellum is one of the two recognised thermophilic *Clostridia* species, the other being *C. stercorarium* [247]. This latter organism grows optimally at 65 °C and its endoglucanase has a half-life of 1 h at 85 °C [248]. Other reports of extremely thermophilic cellulolytic organisms are limited. Ljungdahl et al. [249] gave a preliminary account of some enrichment cultures growing on cellulose at up to 84 °C. In this laboratory studies on several extremely thermophilic isolates with cellulolytic activity have been undertaken [250–255]. The isolates grown at 75 °C were sufficiently different from known cellulolytic anaerobes to warrant their classification into at least one new genus, with the name "*Caldocellum saccharolyticum*" proposed for one of these [253]. For this organism, carboxyme- thylcellulase (CMCase) activity and stability varied between isolates, the most stable having a half-life in excess of 10 h at 85 °C [250, 252]. Certain of the cellulase and hemicellulase (including CMCase, Avicelase, β-glucosidase, xylanase, mann- anase and β-xylosidase) components of "*Caldocellum saccharolyticum*" have been purified [256–259] from *E. coli* clones (see below). The same group of workers has also characterised the β-glucosidase from another cellulolytic organism [260], the enzyme having broad specificity, hydrolysing sophorose and lactose, as well as cellobiose, at appreciable rates. This type of enzyme catalyses the final saccharifica- tion of cellulose and is therefore useful in industrial processes of this type. Other studies on the β-glucosidases from extreme thermophiles include those on a *Thermus* enzyme [261, 262] and on the enzymes from several archaebacterial and eubacterial strains growing at 75 °C or above [227, 228]. Thermophilic β- glucosidases have also been shown in the moderate thermophile *Clostridium thermocellum* [263, 264].

Xylanases have been reported in two extremely thermophilic *Bacillus* strains growing at 65 °C [265, 266]. The enzyme from the *B. stearothermophilus* variant [265] was induced by xylan in the growth medium and degraded the substrate to about 45% in 5 h at 80 °C. The enzyme was maximally active at neutral pH and at 80 °C in 25 min assays. The second *Bacillus* xylanase was from an acidophilic variant [266] grown optimally at 65 °C and pH 3.5–4.0. The inducible enzyme gave maximum activity at 80 °C in 10 min assays at pH 4.0. Both enzymes are considerably more stable than that from a moderately thermophilic *Thermo- monospora* strain [267]. Work on the xylanase from a *Thermotoga* strain [227, 228, 268], however, indicates that this enzyme is the most thermostable xylanase reported to date.

β-Xylosidases catalyse the hydrolysis of xylo-oligosaccharides to xylose and are thus important for the final stages of saccharification of xylan. Few reports of ther- mostable xylosidases are available and these are not, in the main, from extreme ther- mophiles [269, 270]. An aryl β-xylosidase cloned from "*C. saccharolyticum*" into *E. coli* has, however, been characterised [257, 259]. The enzyme had a pI of 4.3 and a pH optimum of 5.7. Activity was specific to *o*- and *p*-nitrophenyl β-D- xylopyranosides; activity against *p*-nitrophenyl α-L-arabinoside was minimal whilst a range of other nitrophenyl-linked sugars and xylobiose were not hydrolysed. Activity against xylan was extremely low, but the xylanase-induced release of reducing sugars from xylan increased by 30% when the enzyme was included in such incubations. As a homogeneously pure preparation the aryl β-xylosidase was only moderately thermostable. Stability was improved by the addition of BSA or dithiothreitol, but not to the level of that given by a crude extract of the enzyme.

In dairy products, lactose can be undesirable due to problems with lactose intolerance and β-galactosidases are therefore of commercial interest for processing lactose-rich dairy products. Several of the β-glucosidases reported above have good β-galactosidase activity. Eubacterial sources of β-galactosidase include

Thermus strain 4.1 A [271], a *T. aquaticus* isolate [272] and an anaerobic *Thermoanaerobacter* strain [273].

Archaebacterial β-galactosidases include that from *Caldariella acidophila* (i.e. *Sulfolobus solfataricus*) with an optimum pH of 5.0 [274]. The stability properties of the enzyme were retained on immobilization of the organism in polyacrylamide gel [275]. In a recent paper [276] describing the properties of a more highly purified preparation of the enzyme, however, the pH optimum of the enzyme was determined to be pH 6.5 (in sodium phosphate buffer, pH measured at room temperature) and its thermostability was apparently some twenty times lower at 85 °C than the more crude preparation or the immobilized enzyme. Two other archaebacterial β-galactosidases, one from the *Desulfurococcus* strain Tok 12 S.1 and the other from *Thermococcus* strain AN1 have been shown to be extremely thermostable; the former being a possible two component system and the latter a single enzyme [277].

4.3.3 Peptide Hydrolases

Proteases account for the majority of enzyme sales in industry, the main markets being in the cleaning and food industries, with a small proportion in the chemical and pharmaceutical industries. The scope for extremely (thermo)stable proteases in these commercial niches has been commented on in several reviews [e.g. 278]. Because of the denaturing effect of heat, when mesophilic proteins are substrates for proteases from extreme thermophiles very high specific activities can be attained at high temperatures. Temperature coefficients are also high.

Several groups of workers have studied proteases from moderately thermophilic *Bacillus* strains, and thermolysin, from *B. thermoproteolyticus*, has been the subject of intensive research [279]. Heinen and Heinen [280] presented one of the earliest reports on the production of an extracellular protease from a *Bacillus* strain growing at 72 °C.

A number of extremely thermophilic *Bacillus* strains grown at 65 °C have been screened for extracellular proteases in this laboratory and the enzymes from two particular strains studied in some detail [281, 282]. The enzyme from *Bacillus* strain EA.1 was a neutral, 42 kDa metalloproteinase susceptible to *o*-phenanthroline inhibition and stabilized by calcium. The second enzyme, from *Bacillus* strain OK3A.1 was also a neutral metalloproteinase but was smaller (32 kDa), had a broader pH optimum, was resistant to *o*-phenanthroline but was less stable. Neither of these two *Bacillus* enzymes showed elastolytic activity, nor activity against synthetic peptide substrates often used for assay of metalloproteinase activity. An asporogenous mutant of *B. stearothermophilus* strain NLI-308-1 has been shown to produce 4-fold greater levels of a metalloproteinase than did the wild-type strain, the enzyme having a probable requirement for zinc and calcium for activity and stability respectively [283]. In 30 min assays at pH 7.0 the enzyme gave maximum activity at 80 °C in the presence of 10 mM $CaCl_2$. The enzyme had a probable molecular weight of 37000 and antibody studies showed that it shared common antigenic determinants with thermolysin. Further, insulin hydrolysis patterns (on oxidised β-chain) for thermolysin and the mutant *Bacillus* enzyme were virtually identical, indicating similar peptide bond specificities.

Caldolysin is the trivial name of the serine proteinase from *T. aquaticus* strain T351 [284]. This enzyme did not have detectable esterase activity and hydrolysis of small peptides of less than four amino acids was not observed. The enzyme was highly stable in the presence of calcium ions. Caldolysin bound six calcium ions per molecule of enzyme, there being both high and low affinity binding sites [285]. Stability of apocaldolysin (i.e. caldolysin treated with EDTA to remove all calcium ions) was restored upon incubation with either calcium or lanthanide ions, the latter giving a lanthanide-caldolysin complex more stable than the native enzyme. Strontium ions were the only other divalent metal ions tested that could restore more than 50% activity.

On immobilization, caldolysin presented enhanced thermostability with sepharose 4B and CM-cellulose as supports (approx. 3 and 2-fold respectively at 90 °C) but decreased thermostability (by a factor of 2 at 90 °C) was observed using controlled pore glass [286]. Recovery of activity after immobilization was as low as 30% or 10% on CM-cellulose and controlled pore glass, perhaps due to steric hindrance.

The serine proteases from three other strains of *T. aquaticus* have been compared to caldolysin [287]. The enzymes had similar pH optima in the range 8.8–9.6 with casein as substrate, there being greater variation between the enzymes using the azodye-linked derivative azo-casein. Caldolysin was the only one of the four enzymes found to be significantly affected by chelators, due to loss of calcium-mediated stability. The specific activities of the enzymes were all found to be higher than their counterparts in mesophilic organisms, a consequence of thermally induced changes in substrate susceptibility to hydrolysis [288].

Proteases have also been described from other *Thermus* isolates. *T. aquaticus* YT-1 was found to produce at least two kinds of extracellular proteinases, termed aqualysins I and II [289]. The former was found to have a pH optimum of 10.4, the latter a pH optimum of 7.0. Both enzymes were inhibited by diisopropylfluorophosphate, a serine protease inhibitor, but EDTA was also effective on the neutral protease. Assay temperatures used for the two enzymes were 70 °C and 95 °C for aqualysins I and II respectively, these being used as "optimum" temperatures, the time of assay being unspecified. Aqualysin I was subsequently found to be stabilized by both calcium ions (maximum activity then being given at 80 °C), and strontium ions [290]. Data on aqualysin II are not yet available. Recent studies on aqualysin I suggest that the enzyme has a precursor form comprising an N-terminal prepro-sequence, the active protease and a C-terminal pro-sequence [291].

The serine protease from *"Thermus caldophilus"* strain GK-24 [292] gave maximum activity at 90 °C in 20 min assays and had a broad pH optimum with casein as substrate. The enzyme showed hydrolytic activity on some small peptide substrates (e.g. CBZ-L-leu-L-tyr-NH$_2$) and also possessed esterase activity. Hydrolysis of synthetic chromogenic peptides and esters is also a property of the recently described serine protease, caldolase, from *Thermus* strain ToK3 [293]. This enzyme contained 10% carbohydrate and four disulphide bonds, but neither calcium nor zinc were detected in the purified enzyme. Thermostability of the enzyme was high in 0.4 M NaCl, but in low ionic strength buffer rapid thermal denaturation occurred at 75 °C. Work has also been undertaken in this laboratory on the extracellular serine proteinase from *Thermus* strain Rt4.1A (Peek K et al.: unpublished observations). The purified enzyme, 34 kDa, contained two disulphide bonds, was stabilized by calcium ions and exhibited maximum activity at 90 °C in 10 min assays (pH 8.0). The enzyme cleaved some synthetic amino acid *p*-nitrophenol esters (especially that of tyrosine) and certain synthetic peptide *p*-nitroanilides, but with distinct specificity differences from aqualysin I [291] and caldolase [294]. Specificity studies on oxidized insulin β-chain were also undertaken and showed that the initial cleavage site was at Leu^{15}-Tyr^{16}. Studies have also been made on the effects of media composition on the stability and yield of the proteinase [294].

The protease (archaelysin) from the archaebacterial *Desulfurococcus* strain Tok 12S.1 growing anaerobically at 88 °C [295] was found to be a serine protease, being inhibited by phenylmethylsulfonyl fluoride and other serine protease inhibitors. The enzyme was markedly different from classic serine proteases, such as trypsin and chymotrypsin, in its lack of reaction to tosyl-L-lysyl-(or phenyl-alanyl-)chloromethane. Specificity studies indicated that the enzyme had a preference for hydrophobic residues on the C-terminal side of the hydrolysis site. The enzyme had a pH optimum of 7.2 and was sensitive to oxidising agents but resistant to certain detergents and chaotrophic agents. Thermostability, which was not dependent upon calcium ions, was greater than other known proteases but was affected by storage (in solution) at low temperature. This observation was explained by Cowan et al. [295] as being a result of extremely limited autolytic cleavage. The proteases from two other strains, *Thermococcus celer* and *Sulfolobus solfataricus* have been shown to possess similar thermostabilities to archaelysin in preliminary investigations [227, 228]. A serine-type proteinase from the archaebacterium *Pyrococcus furiosus* has also been studied, this being a cell envelope associated enzyme inhibited by phenylmethylsulfonylfluoride but resistant to chelators and cysteine-proteinase inhibitors [296]. The enzyme gave multiple active bands on SDS-polyacrylamide gel electrophoresis of 65–140 kDa, but the exact relationships of these active forms are unknown. The enzyme gave "maximum" activity at 115 °C.

An archaebacterial acid protease (optimum pH 2.0) which may represent a new class of enzyme has been purified recently from *Sulfolobus acidocaldarius* [297, 298]. Although the specificity of the enzyme was similar to pepsin, with preference shown towards bonds between large hydrophobic residues, inhibitor studies indicated differences in active site composition. Furthermore, the nucleotide sequence of the enzyme gene was completely dissimilar to those for enzymes of the pepsin family and to pepstatin-insensitive acid protease.

It is worth noting here that modification of α-chymotrypsin by hydrophilization of lysine residues increased the thermostability of the enzyme to a level comparable to the archaebacterial proteases mentioned above [59].

Few other peptide hydrolases from thermophiles have been purified. Cho et al. [299] in studying the aminoacylase from the moderate thermophile *Bacillus thermoglucosidus* found it to have hydrolytic activity against several dipeptides and mention a dipeptidase with high activity against L-valyl-L-alanine, but no characterization was undertaken. Subsequent screening of *Bacillus* species for

dipeptidases by the same group led to the characterization of the enzyme from a
B. stearothermophilus strain [300]. Minagawa [301] reported on the properties of
an aminopeptidase from *Thermus aquaticus* YT-1. The enzyme was a dimeric
metallo-enzyme, inhibited by EDTA and 1,10-phenanthroline. It had a pH
optimum of 8.5–9.0 with fairly broad specificity, albeit with a preference for the
hydrophobic amino acids (the notable exception being proline residues). An
archaebacterial aminopeptidase from *Sulfolobus solfataricus* was identified as a
tetramer with 80 kDa subunits [302]. The enzyme was a metalloexopeptidase, being
inhibited by chelating agents but not by serine protease inhibitors. It had a pH
optimum of 6.5 and in 6 min assays maximum activity was given at 75 °C. Stability
was significantly increased by Co^{2+}, this ion also restoring activity to EDTA-
inhibited enzyme, whereas Ca^{2+}, Mg^{2+} and Zn^{2+} had little effect. The studies of
Minagawa [301] also included data on the carboxypeptidase from *Thermus
aquaticus* YT-1. This was a monomeric enzyme with a similar pH optimum to
the aminopeptidase from the same organism. Specificity data for this enzyme were
limited; the enzyme could cleave off a range of residues, again with the exception
of proline. These types of enzyme are of interest for the debittering of protein
hydrolysates used in the food industry.

4.4 Lyases, Isomerases and Ligases

Lyases described in thermophilic bacteria include aldolase and citrate synthase.
Freeze and Brock [303] reported on the aldolase in *Thermus aquaticus* from an
evolutionary perspective. Enzyme activity was approximately doubled in the
presence of manganese, nickel or zinc at low concentrations and EDTA inhibited
the enzyme at micromolar concentrations.

As with aldolase, citrate synthases have been studied in extreme thermophiles,
specifically archaebacteria, from a taxonomic perspective [304–306]. The enzymes
in *Thermoplasma acidophilum* and *Sulfolobus acidocaldarius* were found to resemble
those in Gram-positive eubacteria and in eukaryotes [305]. The citrate synthase
from *S. acidocaldarius* appeared to be the more stable of the two enzymes.
S. solfataricus, whilst grown at some 5–10 °C higher than the *S. acidocaldarius*
strain, contained a citrate synthetase similar in terms of stability [307]. The citrate
synthase from *T. acidophilum* has recently been cloned [306] and was shown to
have less than 30% similarity overall to the corresponding eubacterial and
eukaryotic enzymes although functionally important sequences were conserved.
S. solfataricus fumarate hydratase has been recently shown to be a tetrameric
170 kDa enzyme with a pH optimum of 8.0 [308]. The enzyme gave maximum
fumarate yield at 85 °C in 10 min assays and was reasonably stable for long periods
at room temperature in methanol, ethanol and propan-2-ol (all at 50%), although
activity was impaired.

Early studies on *Thermus* included the isolation of threonine deaminase from
the strain X-1, this enzyme proving to have a pH optimum of 8.0 and giving
maximum activity at 85 °C to 90 °C in 10 min assays [309].

With regard to isomerases, archaebacteria appear to be unique in their
elaboration of reverse gyrase, the enzyme responsible for imposing positive

supercoiling on closed circular DNA. Whilst eukaryotic and eubacterial cells contain exclusively negatively supercoiled closed circular DNA, it has been demonstrated that the archaebacterium *Sulfolobus acidocaldarius* contains a virus particle (SSV1), the genome of which is positively supercoiled [310]. Reverse gyrases capable of producing positive supercoils in both negatively supercoiled and relaxed closed circular DNA have been shown to be present in not only *S. acidocaldarius* [311–313] but also in thermoacidophilic archaebacteria in general [314]. The specificity of the reverse gyrase has been shown to be similar to that of type I isomerases from eubacteria [315, 316].

Xylose isomerase (glucose isomerase) is an important enzyme in the manufacture of high fructose syrups as well as in xylan degradation since D-xylose can be metabolised via the pentosephosphate pathway after conversion to D-xyluose (and subsequent phosphorylation by D-xylulokinase). The xylose isomerase from *Thermus aquaticus* HB8 has been recently identified as a tetrameric enzyme with identical 50 kDa subunits [317]. The enzyme had a broad pH optimum of 5.5–8.5 for both activity and stability. Maximum stability was conferred by Mn^{2+} or Co^{2+}, but not Mg^{2+}. Immobilization of the enzyme onto epoxy-activated agarose resulted in stability comparable to that of the ion-stabilized free enzyme, but inclusion into nylon beads or hollow fibre entrapment were not as effective [317].

Ligases catalysing carbon-oxygen bond formation are mainly aminoacyl tRNA ligases (synthetases). The tyrosine-tRNA ligase from *"B. caldotenax"* was found to have an amino acid sequence differing in only four residues from the enzyme from other *B. stearothermophilus* strains growing optimally 20 °C below that for *"B. caldotenax"* [318], and the enzymes from the two organisms had identical thermostabilities. An affinity chromatography method for the purification of tyrosine tRNA ligase has been reported [319]. *B. stearothermophilus* arginyl-tRNA synthetase has also been studied [35], thermostability of the enzyme being enhanced by ATP and tRNA. Other aminoacyl-tRNA ligases studied in thermophilic bacteria include the valine- [320], glutamate- [321] and phenylalanine-tRNA ligases [322] and recently crystals of both the seryl- [323] and the threonyl- [324] tRNA ligases from *Thermus thermophilus* have been prepared.

DNA ligase has been purified from *Thermus thermophilus* HB8 [325], the enzyme being NAD^+-dependent, requiring either Mg^{2+} or Mn^{2+} for activity (the latter being preferred) and activated by K^+ and NH_4^+. At the optimum pH of 7.6, the enzyme displayed maximum nick-closing activity in 30 min assays at temperatures in the range of 65–72 °C. Ligation activity of the enzyme was subsequently shown to correlate with nick-closing activity in terms of temperature requirements in the presence of polyethylene glycol [326].

Of the other ligase enzymes, only glutamine synthetase (i.e. glutamate-ammonia ligase) has been studied in any detail. *"B. caldolyticus"* contains two glutamine synthetases, designated E-I and E-II, the latter being the more intrinsically stable, but in the presence of a combination of Mn^{2+}, ATP and L-glutamate both enzymes had similar stability [77]. Glutamine synthetase from a *Bacillus stearothermophilus* strain grown at 65 °C was found to be most active between 70 °C and 75 °C, depending on which divalent metal ion was included in the incubation [33]. Stability of the enzyme was also partly ion dependent.

5 Genetic Engineering

Because of the relative instability of the host enzymes, the cloning of enzymes from extreme thermophiles into an appropriate mesophilic host offers the prospect of a rapid and simple purification step using heat (e.g. β-glucosidase [257], DNA polymerase [327]), with the simultaneous complete inactivation of any contaminating or interfering enzymes. Given the expense of purifying industrial enzymes, and the barrier this has raised to the industrial use of intracellular enzymes, cloning has important implications for the potential commercial use of enzymes from extreme thermophiles.

Relatively little work has been undertaken on either the genetics or the cloning of genes from extreme thermophiles. For a review on the genetics of thermophiles and extreme thermophiles the reader is referred to the review by Bergquist et al. [328].

The malate dehydrogenase gene from *"Thermus flavus"* has been cloned into *E. coli*, the subsequent purification of the gene product from the clone involving a heat treatment step at 88 °C for 15 min. This step was used after disruption of the cells by sonication, ammonium sulphate fractionation and dye-ligand column chromatography [120]. The malate dehydrogenase produced by the clone was found to be virtually identical to the native enzyme in terms of thermostability, Km values for NADH and oxaloacetate and immunological properties. The expression of the gene was low, however, the specific activity of the malate dehydrogenase in the *E. coli* clone being about 30 times lower than in the native organism. Similarly, the 3-isopropylmalate dehydrogenase gene from *Thermus thermophilus* HB8 has been cloned into *E. coli* [137, 329]. Heat treatment at 70 °C for 10 min resulted in a 4-fold purification of the enzyme with 100% yield and although thermostability tests were not performed, maximum activity was given by both the native and the cloned enzyme in 20 min assays at the same temperature, even after repeated subculturing of the mesophilic clone [137]. The glyceraldehyde-3-phosphate dehydrogenase from the archaebacterium *Pyrococcus woesei* has also been expressed at a high level in *E. coli* [147]. Purification of the cloned enzyme was through a heat incubation followed by affinity chromatography. The enzyme proved to have virtually identical kinetic properties and stability (intrinsic or enhanced by salts) to the native enzyme.

The endo-β-glucanase gene from *Thermoanaerobacter cellulolyticus* [330] has been cloned into *E. coli* [331] and recloned into *Saccharomyces cerevisiae* [332]. The thermostability of the enzyme was similar in the native organism and in the *E. coli* clone, no data being given for the enzyme from the *S. cerevisiae* clone. The enzyme was not secreted from the *E. coli* clone but was reported as being retained in the cytoplasm [331]. The carboxymethylcellulase activity from *Clostridium thermocellum* was similarly contained within an *E. coli* clone, but when the gene encoding for the enzyme was further transferred into *Bacillus* species the gene product was secreted [333].

A number of cellulase and hemicellulase genes have been cloned from *"Caldocellum saccharolyticum"* into *E. coli*, including Avicelase, CMCase, β-glucosidase, xylanase, xylosidase and mannanase by Bergquist's group [328, 334–337]. The thermal stabilities of these have been unaffected, and heat treatment shown to give 5 to 20-fold purification with recoveries usually better than 75% [257]. The Avicelase is unusual in that it is a protein of moderate molecular weight (84 kDa) capable of degrading crystalline cellulose [258]. It has significant activity against xylan and CMC. The CMCase may have more than one active site, with two pH optima and a downward deflection of the Lineweaver Burk plot at high substrate concentration [338]. The β-glucosidase has been well characterised [256]. It has broad specificity for β-D-glucosides, galactosides, fucosides and xylosides. All three of these enzymes have significant half-lives at 90 °C (Ruttersmith, pers. comm. and [256, 338]). The xylanase is endo-acting with no activity against xylobiose (Schofield L R, Daniel R M: unpublished observations). The xylosidase is also inactive against xylobiose, and has significant activity only against p-nitrophenol arabinopyranoside [259]. Luthi et al. [339, 340] have achieved over-expression of both the xylanase and the acetyl xylan esterase from "Caldocellum saccharolyticum" in E. coli. In the case of the xylanase the enzyme comprised 20% of the total cell protein, whilst for the acetyl xylan esterase the level was 10%.

The pullulanase gene from the anaerobe Thermoanaerobium brockii has been cloned into both E. coli and B. subtilis [341]. The former clone did not secrete the enzyme whereas the latter secreted more activity than did the native organism. The enzyme secreted by B. subtilis was not however, glycosylated (in contrast to the native enzyme) and exhibited impaired thermostability, the half-life at 70 °C being less than half that of the native enzyme. Both the α- and β-galactosidases from Thermus strain T2 have been cloned not only into E. coli but also into T. thermophilus HB27 by Koyama et al. [342]. This same group has used a direct plasmid transfer method to clone T. thermophilus HB27 tryptophan synthetase [343].

Motoshima et al. [344] have achieved a high level of expression in E. coli of the aminopeptidase T from Thermus aquaticus YT-1. 5% of the soluble protein was identified as being the enzyme which could be purified from the mesophilic host using heat treatment. With regard to cloning of other peptide hydrolases, expression of the aqualysin I gene from T. aquaticus YT-1 in E. coli resulted in the accumulation of inactive protease in the membrane [291]. The enzyme could be released in a soluble, active form by heat treatment of the membrane at 65 °C. Further, when the enzyme was mutated to give a Ser-Ala replacement at the active site the effect of the heat treatment was negated. The interpretation was that E. coli peptidase was responsible for signal peptide cleavage, whilst the N- and C-terminal pro-sequences of the enzyme were cleaved by autoproteolysis [291].

6 Conclusion

As this review indicates, a wide variety of stable enzymes are available from extreme thermophiles. Furthermore, the variety of these organisms found over a relatively short time is an indication that the resource has barely been tapped.

Enzymes from extreme thermophiles are not only stable to heat, but also to organic solvents, detergents and chaotropic agents. This opens up the possibility of using enzymes in the presence of denaturing substrates or products, and in harsh conditions generally. Except in the case of enzymes which are characteristically only found in eukaryotes, enzyme stability per se should no longer be a problem in enzyme mediated bio-industrial processes.

Cloned thermostable enzymes offer the early prospect of readily and cheaply purified intracellular enzymes. Only a few percent of the cell mass required for the wild type need be used, assuming over production of the enzyme from the cloned gene, and no contaminating activities will be present.

In the short term there are a number of problems. Relatively little research has been done so that most enzymes and organisms are uncharacterised and lack FDA approval. Running existing processes at higher temperatures will need process redesign. Enzyme yields are of course low since little work has taken place on growth optimisation, let alone strain selection and enzyme yield optimisation; archaebacteria in particular are difficult to grow on a large scale. Given advances in genetic engineering, however, low enzyme yields are unlikely to be an insurmountable problem.

7 References

1. Woese GR, Fox GE (1977) Proc Natl Acad Sci USA 74: 5088–5090
2. Stetter KO, König H, Stackebrandt E (1983) Syst Appl Microbiol 4: 535–551
3. Huber R, Kurr M, Jannash HW, Stetter KO (1989) Nature 342: 833–834
4. Pledger RJ, Baross JA (1991) J Gen Micro 137: 203–211
5. Zillig W, Holz I, Janekovic D, Klenk HP, Imsel E, Trent J, Wonderl S, Forjaz VH, Coutinho R, Ferreira T (1990) J Bacteriol 172: 3959–3965
6. Kandler O, Hippe H (1977) Arch Microbiol 113: 57–60
7. Kandler O, König H (1978) Arch Microbiol 118: 141–152
8. Tornabene TG, Langworthy TA (1979) Science 203: 51–53
9. Tornabene TG, Langworthy TA, Holzer G, Oro J (1979) J Mol Evol 13: 73–83
10. De Rosa M, Gambacorta A, Gliozzi A (1986) Microbiol Rev 50: 70–80
11. Thurl S, Schäfer W (1988) Biochim Biophys Acta 961: 253–261
12. Danson MJ (1988) Adv Microbiol Physiol 29: 165–231
13. Danson MJ (1989) Can J Microbiol 35: 58–64
14. Huber R, Langworthy TA, König H, Thomm M, Woese CR, Sleyter UB, Stetter KO (1986) Arch Microbiol 144: 324–333
15. Huser BA, Patel BKC, Daniel RM, Morgan HW (1986) FEMS Microbiol Lett 37: 121–127
16. Jannasch HW, Huber R, Belkin S, Stetter KO (1988) Arch Microbiol 150: 103–104
17. Achenbach-Richter L, Gupta R, Stetter KO, Woese CR (1987) Syst Appl Microbiol 9: 34–39
18. Sharp RJ, Brown KJ, Atkinson A (1980) J Gen Microbiol 117: 201–210
19. Hudson JA, Morgan HW, Daniel RM (1986) J Gen Microbiol 132: 531–540
20. Hudson JA, Morgan HW, Daniel RM (1987) Syst Appl Microbiol 9: 218–223
21. Donnison AM, Gutteridge CS, Norris JR, Morgan HW, Daniel RM (1986) J Anal Appl Pyrolysis 9: 281–295
22. Hensel R, Demharter W, Kandler O, Kroppenstedt RM, Stackebrant E (1986) Int J Syst Bact 36: 444–453
23. Schofield KM (1988) MSc Thesis, University of Waikato, Hamilton, New Zealand
24. Sharp RJ, Williams RAD (1988) Appl Environ Microbiol 54: 2049–2053
25. Ahern TJ, Klibanov AM (1985) Science 228: 1280–1284

26. Ahern TJ, Klibanov AM (1986) Why do enzymes irreversibly inactivate at high temperature? In: Oxender DL (ed) Protein structure folding and design. Liss, New York, p 283
27. Zale SE, Klibanov AM (1986) Biochemistry 25: 5432–5444
28. Aswad DW (1990) Annals NY Acad Sci 613: 26–36
29. McLinden J, Murdock A, Amelunxen R (1986) Biochim Biophys Acta 871: 207–216
30. Wedler FC, Hoffman FM (1974) Biochemistry 13: 3215–3221
31. Wedler FC, Hoffman FM, Kenny R, Garfi J (1976) Maintenance of specificity, information and thermostability in thermophilic *Bacillus* sp. glutamine synthetase. In: Zuber H (ed) Enzymes and proteins from thermophilic organisms. Birkhäuser Verlag, Basel, p 187
32. Hibino Y, Nosoh Y, Samejima T (1974) J Biochem 75: 553–561
33. Hachimori A, Matsunaga A, Shimizu M, Samejima T, Nosoh Y (1974) Biochim Biophys Acta 350: 461–474
34. Parfait R (1973) FEBS Lett 29: 323–325
35. Brandts JF (1967) Heat effects on proteins and enzymes. In: Rose AH (ed) Thermobiology. Academic, New York, p 25
36. Jaenicke R (1981) Ann Rev Biophys Bioeng 10: 1–67
37. Matthews BW (1987) Biochemistry 26: 6885–6888
38. Daniel RM (1986) The stability of proteins from extreme thermophiles. In: Oxender DL (ed) Protein structure folding and design. Liss, New York, p 291
39. Klibanov A (1983) Adv Appl Microbiol 29: 1–28
40. Langridge J (1968) J Bacteriol 96: 1711–1717
41. Grutter MG, Hawkes RB, Matthews BW (1979) Nature 277: 667–669
42. Alber T (1989) Ann Rev Biochem 58: 765–798
43. Perutz MF, Raidt H (1975) Nature 255: 256–259
44. Yutani K, Ogasahara K, Sugino Y, Matsushiro A (1977) Nature 267: 274–275
45. D'Souza VT, Hanabusa K, O'Leary T, Gadwood RC, Bender ML (1985) Biochem Biophys Res Commun 129: 727–732
46. D'Souza VT, Lu XL, Ginger RD, Bender ML (1987) Proc Natl Acad Sci USA 84: 673–674
47. Mutter M (1985) Angew Chem Int Engl 24: 639–653
48. Ahern TJ, Casal JI, Petsko GA, Klibanov AM (1987) Proc Natl Acad Sci USA 84: 675–679
49. Daniel RM, Cowan DA, Morgan HW, Curran MP (1982) Biochem J 207: 641–644
50. Veronese FM, Boccu E, Schiavon O, Grandi C, Fontana A (1984) J Appl Biochem 6: 39–47
51. Owusu R, Cowan DA (1989) Enzyme Microb Technol 12: 374–377
52. Vihinen M (1987) Prot Eng 1: 477–480
53. Yamamoto K, Nagao T, Makino Y, Urabe I, Okada H (1990) Annals NY Acad Sci 613: 362–365
54. Waldvogel S, Weber H, Zuber H (1987) Biol Chem Hoppe-Seyler 368: 1391–1399
55. Liao MD, McKenzie T, Hageman R (1986) Proc Natl Acd Sci USA 83: 576–580
56. Stellwagen E (1984) Annals NY Acad Sci 434: 1–6
57. Melik-Nubarov NS, Mozhaev VV, Siksnis S, Martinek K (1987) Biotech Lett 9: 725–730
58. Melik-Nubarov NS, Siksnis VA, Slepev VI, Shchegolev AA, Mozhaev VV (1990) Mol Biol (Moscow) 24: 346–357
59. Mozhaev VV, Siksnis VA, Melik-Nubarov NS, Galkantaite NZ, Denis GJ, Butkus EP, Zaslavsky BYu, Mestechkina NM, Martinek K (1988) Eur J Biochem 173: 147–154
60. Wigley DB, Clarke AR, Dunn CR, Barstow DA, Atkinson T, Chia WN, Muirhead H, Holbrook JJ (1987) Biochim Biophys Acta 916: 145–148
61. Qaw FS, Brewer JM (1988) Mol Cell Biochem 71: 121–127
62. Merkler DJ, Farrington GK, Wedler FC (1981) Int J Prot Pept Res 18: 430–432
63. Cupo P, El-Deiry W, Whitney PL, Awad WM Jr (1980) J Biol Chem 255: 10828–10833
64. Fabry S, Lang J, Niermann T, Vingron M, Hensel R (1989) Eur J Biochem 179: 405–414

65. Mendendez-Arias L, Argos P (1989) J Mol Biol 206: 397–406
66. Suzuki Y, Oishi K, Nakano H, Nagayama T (1987) Appl Microbiol Biotechnol 26: 546–551
67. Matthews BW, Nicholson H, Becktel WJ (1987) Proc Natl Acad Sci USA 84: 6663–6667
68. Imanaka T, Shibazaki M, Takagi M (1986) Nature 324: 695–697
69. Querol E, Parrilla A (1987) Enz Microbiol Technol 9: 238–244
70. Schultes V, Deutzmann R, Jaenicke R (1990) Eur J Biochem 191: 25–32
71. Wrba A, Schweiger A, Schultes V, Jaenicke R, Zavdszky P (1990) Biochemistry 29: 7584–7592
72. Biesecker G, Harris JI, Thierry JC, Walker JE, Wonacott AJ (1977) Nature 266: 328–333
73. Iijima S, Saiki T, Beppu T (1984) J Biochem 95: 1273–1281
74. Torchilin VP, Maksimenko AV, Smirnov VN, Berezin IV, Klibanov AM, Martinek K (1978) Biochim Biophys Acta 522: 277–283
75. Torchilin VP, Trubetskoi VS, Omel'yanenko VG, Martinek K (1983) J Mol Catal 19: 291–301
76. Trubetskoi VS, Torchilin VP (1985) Int J Biochem 17: 661–663
77. Merkler DJ, Srikumar K, Marchese-Ragona SP, Wedler FC (1988) Biochim Biophys Acta 952: 101–114
78. Wedler FC, Merkler DJ (1985) Curr Topics Cell Reg 26: 263–280
79. Dixon M, Webb EG (1979) Enzymes. Longman Group Ltd., London, pp 47–206
80. Silvius JR, Read BD, McElhabey RN (1978) Science 199: 902–904
81. Kubo K (1985) J Theor Biol 115: 551–559
82. Wolf J, Bagnall D (1979) Statistical tests to decide between straight line segments and curves as suitable fits to Arrhenius plots or other data. In: Lyons JM, Graham D, Raison JK (eds) Low temperature stress in crop plants: the role of the membrane. Academic, New York, p 528
83. Hickey CW, Daniel RM (1979) J Gen Microbiol 114: 195–200
84. Mäntsälä P (1985) Biochem Int 10: 955–962
85. Doig AR Jr (1974) Stabilization of enzymes from thermophilic microorganisms. In: Pye EK, Wingard LB Jr (eds) Enzyme engineering 2. Plenum, New York, p 17
86. Sonnleitner B, Fiechter A (1983) TIBTech 1: 74–80
87. Hartley BS, Payton MA (1983) Biochem Soc Symp 48: 133–146
88. Wiegel J, Ljungdahl LG (1986) CRC Critical Reviews in Biotechnology 3: 39–108
89. Daniel RM, Cowan DA, Morgan HW (1981) Chemistry in New Zealand (1981) *June*: 94–97
90. Daniel RM, Morgan HW, Martin AM (1986) Industrial Biotechnology (1986) *Jan*: 89–91
91. Slapack GE, Russell I, Stewart GG (1987) Thermophilic Microbes in Ethanol Production. CRC Press, Florida, USA
92. Ward OP, Moo-Young M (1988) Biotech Adv 6: 39–69
93. Daniel RM (1986) J Theor Biol 120: 125–127
94. Legoy MD, Kim HS, Thomas D (1985) Process Biochem 20: 145–148
95. Hayakawa K, Urabe I, Okada H (1985) J Ferment Technol 63: 245–250
96. Cocco D, Rinaldi A, Savini I, Cooper JM, Bannister JV (1988) Eur J Biochem 174: 267–271
97. Nagata S, Feicht R, Bette W, Gunther H, Simon H (1987) Appl Microbiol Biotechnol 26: 263–267
98. Keinan E, Hafeli EK, Seth KK, Lamed R (1986) J Am Chem Soc 108: 162–169
99. Keinan E, Seth KK, Lamed R (1986) J Am Chem Soc 108: 3474–3480
100. Rella R, Raia CA, Pensa M, Pisani FM, Gambacorta A, De Rosa M, Rossi M (1987) Eur J Biochem 167: 475–479
101. Bryant F, Ljungdahl LG (1981) Biochem Biophys Res Commun 100: 793–799
102. Bryant FO, Wiegel J, Ljungdahl LG (1988) Appl Environ Microbiol 54: 460–465
103. Lamed RJ, Zeikus JG (1981) Biochem J 195: 183–190
104. Lamed RJ, Keinan E, Zeikus JG (1981) Enzyme Microb Technol 3: 144–148

105. Taguchi H, Yamashita M, Matsuzawa H, Ohta T (1982) J Biochem 91: 1343–1348
106. Taguchi H, Machida M, Matsuzawa H, Ohta T (1985) Agric Biol Chem 49: 359–365
107. Schär H-P, Zuber H (1979) Hoppe-Seyler's Z Physiol Chem 360: 796–807
108. Tratschin JD, Wirz B, Frank G, Zuber H (1983) Hoppe-Seyler's Z Physiol Chem 364: 879–892
109. Wirz B, Suter F, Zuber H (1983) Hoppe-Seyler's Z Physiol Chem 364: 893–909
110. Clarke AR, Waldman ADB, Munro I, Holbrook JJ (1985) Biochim Biophys Acta 828: 375–396
111. Clarke AR, Atkinson T, Campbell JW, Holbrook JJ (1985) Biochim Biophys Acta 829: 387–396
112. Clarke AR, Evington JRN, Dunn CR, Atkinson T, Holbrook JJ (1986) Biochim Biophys Acta 870: 112–126
113. Turunen M, Parkkinen E, Londesborough J, Korkola M (1987) J Gen Microbiol 133: 2865–2873
114. Wrba A, Jaenicke R, Huber R, Stetter KO (1990) Eur J Biochem 188: 195–201
115. Bastow DA, Clarke AR, Chin WN, Wigley D, Sharman AF, Holbrook JJ, Atkinson T, Minton NP (1986) Gene 46: 47–55
116. Clarke AR, Wigley DB, Barstow DA, Chin WN, Atkinson T, Holbrook JJ (1987) Biochim Biophys Acta 913: 72–80
117. Iijima S, Saiki T, Beppu T (1980) Biochim Biophys. Acta 613: 1–9
118. Iijima S, Oh MJ, Saiki T, Beppu T (1981) Agric Biol Chem 45: 773–774
119. Iijima S, Oh MJ, Saiki T, Beppu T (1986) J Biochem 99: 1667–1672
120. Iijima S, Uozumi T, Beppu T (1986) Agric Biol Chem 50: 589–592
121. Smith K, Sundaram TK (1988) Biochim Biophys Acta 955: 203–213
122. Pask-Hughes RA, Williams RAD (1977) J Gen Microbiol 102: 375–383
123. Murphey WH, Kitto GB, Everse J, Kaplan NO (1967) Biochemistry 6: 603–609
124. Grossebüter W, Hartl T, Görisch H, Stetzowski JJ (1986) Biol Chem Hoppe-Seyler 367: 457–463
125. Hartl T, Grossebüter W, Görisch H, Stetzowski JJ (1987) Biol Chem Hoppe-Seyler 368: 259–267
126. Görisch H, Hartl T, Grossebüter W, Stetzowski JJ (1985) Biochem J 226: 885–888
127. Honka E, Fabry S, Niermann T, Palm, P, Hensel R (1990) Eur J Biochem 188: 623–632
128. Hensel R, König H (1988) FEMS Microbiol Lett 49: 75–79
129. Bugden N, Danson MJ (1986) FEBS Lett 196: 207–210
130. Zeikus JG, Fuchs G, Kenealy K, Thauer RK (1977) J Bacteriol 132: 604–613
131. Danson MJ, Wood PA (1984) FEBS Lett 172: 289–293
132. Bartolucci S, Rella R, Guagliardi A, Raia CA, Gambacorta A, De Rosa M, Rossi M (1987) J Biol Chem 262: 7725–7731
133. Guagliardi A, Manco G, Rossi M, Bartolucci S (1989) Eur J Biochem 183: 25–30
134. Guagliardi A, Raia CA, Rella R, Bückmann AF, D'Auria S, Rossi M, Bartolucci S (1991) Biotechnol Appl Biochem 13: 25–35
135. Ramaley RF, Hudock MO (1973) Biochim Biophys Acta 315: 22–36
136. Eguchi H, Wakagi T, Oshima T (1989) Biochim Biophys Acta 990: 133–137
137. Yamada T, Akutsu N, Miyazaki K, Kakinuma K, Yoshida M, Oshima T (1990) J Biochem 108: 449–456
138. Vali Z, Kilar F, Lakatos S, Venyaminor SA, Zavodszky P (1980) Biochim Biophys Acta 615: 34–47
139. Giardina P, De Biasi M-G, De Rosa M, Gambacorta A, Buonocore V (1986) Biochem J 239: 517–522
140. Hocking JD, Harris JI (1973) FEBS Lett 34: 280–284
141. Harris JI, Hocking JD, Runswick MJ, Suzuki K, Walker JE (1980) J Biochem 108: 535–547
142. Fabry S, Hensel R (1987) Eur J Biochem 165: 147–155
143. Hensel R, Laumann S, Lang J, Heumann H, Lottspeich F (1987) Eur J Biochem 170: 325–333
144. Hocking JD, Harris JI (1980) Eur J Biochem 108: 567–579

145. Walker JE, Carne AF, Runswick MJ, Bridgen J, Harris JI (1980) Eur J Biochem 108: 549–565
146. Walker JE, Wonacott AJ, Harris JI (1980) Eur J Biochem 108: 581–586
147. Zwickl P, Fabry S, Bogedain C, Haas A, Hensel R (1990) J Bacteriol 172: 4329–4338
148. Wakao H, Wakagi T, Oshima T (1987) J Biochem 102: 255–262
149. Kawada N, Takeda K, Nosoh Y (1981) J Biochem 89: 1017–1027
150. Walsh KAJ, Daniel RM, Morgan HW (1983) Biochem J 209: 427–433
151. Pinkwart M, Schneider K, Schlegel HG (1983) FEMS Microbiol Lett 17: 137–141
152. Pusheva MA, Savel'eva ND (1982) Mikrobiologiya 51: 896–900 (Eng Trans 703–707)
153. Pusheva MA, Berezu VI, Savel'eva ND, Kryukov VR (1987) Prikl Biokhim Mikrobiol 23: 185–191
154. Shah NN, Clark DS (1990) Appl Environ Microbiol 56: 858–863
155. Bryant FO, Adams MWW (1989) J Biol Chem 264: 5070–5079
156. Saiki RK, Scharf S, Faloona F, Mullis KB, Horn GT, Erlich HA, Arhein N (1985) Science 230: 1350–1354
157. Oste C (1988) Bio/Techniques 6: 162–167
158. Saiki RK, Gelfand DH, Stoffel S, Scharf SJ, Higuchi R, Horn GT, Mullis KB, Erlich HA (1988) Science 239: 487–491
159. Kaledin AS, Slyusarenko AG, Gorodetskii SI (1980) Biokhimiya 45: 494–501
160. Kaledin AS, Slyusarenko AG, Gorodetskii SI (1981) Biokhimiya 46: 1247–1254
161. Kaledin AS, Slyusarenko AG, Gorodetskii SI (1982) Biokhimiya 47: 1515–1521
162. Rüttimann C, Cotoras M, Zaldivar J, Vicuna R (1985) Eur J Biochem 149: 41–46
163. Chien A, Edgar DB, Trela JM (1976) J Bacteriol 127: 1550–1557
164. Simpson HD, Coolbear, T, Vermue M, Daniel RM (1991) Biochem Cell Biol 68: 1292–1296
165. Klimczak LJ, Grummt F, Burger KJ (1986) Biochemistry 25: 4850–4855
166. Sprott GD, Jarrell KF (1981) Can J Microbiol 27: 444–451
167. Rossi M, Rella R, Pensa M, Bartolucci S, De Rosa M, Gambacorta A, Raia CA, Dell'aversano Orabona N (1986) Syst Appl Microbiol 7: 337–341
168. Prangishvili DA (1986) Molekulyarna Biologiya 20: 380–390
169. Klimczak LJ, Grummt F, Burger KJ (1986) Nucl Acids Res 13: 5269–5282
170. Thomm MS, Madon J, Stetter KO (1986) Biol Chem Hoppe-Seyler 367: 473–481
171. Prangishvili D, Zillig W, Gierl A, Biesert L, Holz I (1982) Eur J Biochem 122: 471–477
172. Stetter KO, Winter J, Hartlieb R (1980) Z Bakteriol Abt 1 Orig C 1: 201–214
173. Air GM, Harris JI (1974) FEBS Lett 38: 277–281
174. Fabry M, Sumegi J, Venetainer P (1976) Biochim Biophys Acta 435: 228–235
175. Wnendt S, Hartmann RK, Ulbrich N, Erdmann VA (1990) Eur J Biochem 191: 467–472
176. Morozov IA, Gambaryan AS, Lvova TN, Nedasparov AA, Venkstern TV (1982) Eur J Biochem 129: 429–436
177. Kumagi I, Watanabe K, Oshima T (1980) Proc Natl Acad Sci USA 77: 1922–1926
178. Kumagi I, Watanabe K, Oshima T (1982) J Biol Chem 257: 7388–7395
179. Morozov IA, Gambaryan AS, Venkstern TV (1984) Mol Biol (Moscow) 18: 1363–1368
180. Sauer FD (1986) Biochem Biophys Res Commun. 136: 542–547
181. Cacciapuoti G, Porcelli M, Carteni-farina M, Gambacorta A, Zappia V (1986) Eur J Biochem 161: 263–271
182. Marino G, Nitti G, Arnone MI, Sannia G, Gambacorta A, De Rosa M (1988) J Biol Chem 263: 12305–12309
183. Arnone MI, Cubellis MV, Nitti G, Sannia G, Marino G (1988) Ital J Biochem 37: 347–348
184. Cubellis MV, Rozzo C, Nitti G, Arnone MI, Marino G, Sannia G (1989) Eur J Biochem 186: 375–381
185. Schutten I, Harder W, Dijkhuizen L (1987) Appl Microbiol Biotechnol 27: 292–298
186. Hengartner H, Harris JI (1975) FEBS Lett 55: 282–285
187. Yoshida M (1972) Biochemistry 11: 1087–1093
188. Xu J, Oshima T, Yoshida M (1990) J Mol Biol 215: 597–606
189. Yoshizaki F, Imahori H (1979) Agric Biol Chem 43: 527–536

190. Dahl C, Koch H-G, Keuken O, Trüper HG (1990) FEMS Microbiol Lett 67: 27–32
191. Matsunaga A, Koyama N, Nosoh Y (1974) Arch Biochem Biophys 160: 504–513
192. Owusu RK, Cowan DA (1991) Enzyme Microbiol Technol 13: 158–163
193. Sobek H, Görisch H (1988) Biochem J 250: 453–458
194. Yeh M-F, Trela JM (1976) J Biol Chem 251: 3134–3139
195. Smile DH, Donohue M, Yeh M-F, Kenkle T, Trela JM (1977) J Biol Chem 252: 3399–3401
196. Boleznin MI, Bunina ZF, Mazanova VV, Pachkunov DM, Smolyaninov VV (1987) Prikl Biok Mikrobiol 23: 536–541
197. Yoshida M, Oshima T, Imahori K (1973) J Biochem 74: 1183–1191
198. Sato S, Hutchinson CA, Harris JI (1977) Proc. Natl Acad Sci USA 74: 542–546
199. Bingham AHA, Atkinson T, Sciaky D, Roberts RJ (1978) Nucl. Acids Res 5: 3457–3467
200. Catterall JF, Welker NE (1977) J Bacteriol 129: 1110–1120
201. Prangishvili DA, Vashakidze RP, Chelidze MG, Gabriadze IYu (1985) FEBS Lett 192: 57–60
202. Kaboev OK, Luchkina LA, Kuziakina TI (1985) J Bacteriol 164: 878–881
203. Kaboev OK, Luchkina LA, Akhemdov AT, Bekker ML (1981) FEBS Lett 132: 337–340
204. Guy GR, Daniel RM (1982) Biochem J 203: 787–790
205. Curran MP, Daniel RM, Guy GR, Morgan HW (1985) Arch Biochem Biophys 241: 571–576
206. Patchett ML (1988) Ph D Thesis, University of Waikato, Hamilton, New Zealand
207. Patchett ML, Daniel RM, Morgan HW (1991) Biochim Biophys Acta 1077: 291–298
208. Hachimori A, Takeda A, Kaibuchi M, Ohkawara N, Samejima T (1975) J Biochem 77: 1177–1183
209. Kasho VN, Avaeva SM (1984) Int J Biochem 16: 315–321
210. Verhoeven JA, Schenck KM, Meyer RR, Trela JM (1986) J Bacteriol 168: 318–321
211. Wakagi T, Oshima T (1985) Biochim Biophys Acta 817: 33–41
212. Gogarten JP, Kibak H, Dittrich P, Taiz L, Bowman EJ, Bowman BJ, Manolson MF, Poole RJ, Date T, Oshima T, Konisi J, Denda K, Yoshida M (1989) Proc Natl Acad Sci USA 86: 6661–6665
213. Iwabe N, Kuwa K, Hasegawa M, Osawa S, Miyata T (1989) Proc Natl Acad Sci USA 86: 9355–9359
214. Searcy DG, Whatley FR (1982) Z Bakt Mik Hyg Abl 1 Orig C3: 245–257
215. Konishi J, Wakagi T, Oshima T, Yoshida M (1987) J Biochem 102: 1379–1387
216. Lübben M, Schäfer G (1987) Eur J Biochem 164: 533–540
217. Yokoyama K, Oshima T, Yoshida M (1990) J Biol Chem 265: 21946–21950
218. Aunstrup K (1979) Production, isolation and economics of extracellular enzymes. In: Wingood LB Jr, Katchalski-Katzir E, Goldstein L (eds) Applied biochemistry and bioengineering vol 2. Academic, London, p 1
219. Kindle KL (1983) Appl Biochem Biophys 8: 153–170
220. Fogarty WM (1983) Microbial amylases. In: Fogarty WM (ed) Microbial Enzymes and Biotechnology. Applied Science Publishers, London, p 1
221. Saha BC, Zeikus JG (1987) Proc Biochem June: 78–82
222. Slominska L, Starogardzka G (1986) Starch/Stärke 38: 205–210
223. Antranikian G (1990) FEMS Microbiol Rev 75: 201–218
224. Grueninger H, Sonnleitner B, Fiechter A (1984) Appl Microbiol Biotechnol 19: 414–421
225. Uchino F (1982) Agric Biol Chem 46: 7–13
226. Plant AR, Patel BKC, Morgan HW, Daniel RM (1987) Syst Appl Microbiol 9: 158–162
227. Bragger JM, Daniel RM, Coolbear T, Morgan HW (1989) Appl Microbiol Biotech 31: 556–561
228. Daniel RM, Bragger J, Morgan HW (1989) Enzymes from extreme thermophiles. In. Abramowicz DA (ed) Biocatalysis. Van Nostran Reinhold, New York, p 243
229. Koch R, Zablowski P, Spreinat A, Antranikian G (1990) FEMS Microbiol Rev 71: 21–26
230. Giblin M, Kelly CT, Fogarty WM (1987) Can J Microbiol 33: 614–618

231. Plant AR, Clemens RM, Daniel RM, Morgan HW (1987) Appl Microbiol Biotech 26: 427–433
232. Plant AR, Clemens RM, Morgan HW, Daniel RM (1987) Biochem J 246: 537–541
233. Hyun HH, Zeikus JG (1985) Appl Environ Microbiol 49: 1168–1173
234. Suzuki Y, Imai T (1985) Appl Microbiol Biotechnol 21: 20–26
235. Plant AR, Morgan HW, Daniel RM (1986) Enzyme Microb Technol 8: 668–672
236. Melasniemi H (1987) Biochem J 246: 193–197
237. Melasniemi H (1988) Biochem J 250: 813–818
238. Saha BC, Zeikus JG (1989) TIBTech 7: 234–239
239. Suzuki Y, Fujii H, Uemura H, Suzuki M (1987) Starch/Stärke 39: 17–23
240. Plant AR, Parratt S, Daniel RM, Morgan HW (1988) Biochem J 255: 865–868
241. Suzuki Y, Yuki T, Kishigami T, Abe S (1976) Biochim Biophys Acta 445: 386–397
242. Suzuki Y, Ueda Y, Nakamura N, Abe S (1979) Biochim Biophys Acta 566: 62–66
243. Coughlan MP, Folan MA (1979) Int J Biochem 10: 103–108
244. Margaritis A, Merchant RFJ (1986) CRC Critical Reviews in Biotechnology 4: 327–367
245. Weimer PJ (1986) Use of thermophiles for the production of fuels and chemicals. In: Brock TD (ed) Thermophiles: General, molecular and applied microbiology. J Wiley, New York, p 217
246. Marsden WL, Gray PP (1986) CRC Critical Reviews in Biotechnology 3: 235–276
247. Madden RH (1983) Int J Syst Bact 33: 837–840
248. Creuzet N, Frixon C (1983) Biochimie 65: 149–156
249. Ljungdahl LG, Bryant F, Carriera L, Saki T, Wiegel J (1981) Some aspects of thermophilic and extreme thermophilic anaerobic microorganisms. In: Hollaender A (ed) Trends in the biology of fermentation for fuels and chemicals. Plenum, New York, p 397
250. Sharrock KR (1985) PhD Thesis, University of Waikato, Hamilton, New Zealand
251. Reynolds PHS, Sissons CH, Daniel RM, Morgan HW (1986) Appl Environ Microbiol 51: 12–17
252. Sissons CH, Sharrock KR, Daniel RM, Morgan HW (1987) Appl Environ Microbiol 53: 832–838
253. Donnison AM, Brockelsby CM, Daniel RM, Morgan HW (1989) Biotechnol Bioeng 33: 1495–1499
254. Hudson JA, Morgan HW, Daniel RM (1990) Appl Microb Biotech 33: 687–691
255. Hudson JA, Morgan HW, Daniel RM (1991) Appl Microb Biotech 35: 270–273
256. Plant AR, Oliver JE, Patchett ML, Daniel RM, Morgan HW (1988) Arch Biochem Biophys 262: 181–188
257. Patchett ML, Neal TL, Schofield LR, Strange RC, Daniel RM, Morgan HW (1989) Enzyme Microb Technol 11: 113–115
258. Schofield LR, Neal TL, Patchell ML, Strange RC, Daniel RM, Morgan HW (1989) Annals NY Acad Sci 542: 240–243
259. Hudson RC, Schofield LR, Coolbear T, Daniel RM, Morgan HW (1991) Biochem J 273: 645–650
260. Patchett ML, Daniel RM, Morgan HW (1987) Biochem J 243: 779–787
261. Takase M, Horikoshi K (1988) Appl Microbiol Biotechnol 29: 55–60
262. Takase M, Horikoshi K (1989) Agric Biol Chem 53: 559–560
263. Ait N, Creuzet N, Cattaneo J (1979) Biochem Biophys Res Commun 90: 537–546
264. Ait N, Creuzet N, Cattaneo J (1982) J Gen Microbiol 128: 569–577
265. Grüninger H, Fiechter A (1986) Enzyme Microb Technol 8: 309–314
266. Uchino F, Nakane T (1981) Agric Biol Chem 45: 1121–1127
267. Ristroph DL, Humphrey AE (1985) Biotechnol Bioeng. 27: 832–836
268. Simpson HD, Haufler U, Daniel RM (1991) Biochem J 277: 413–417
269. Ristroph DL, Humphrey AE (1985) Biotechnol Bioeng 27: 909–913
270. Sung NK, Kang IS, Chun HK. Akiba T, Horikoshi K (1987) Korean J Appl Microbiol Bioeng 15: 267–272
271. Cowan DA, Daniel RM, Martin AM, Morgan HW (1984) Biotechnol Bioeng 26: 1141–1145
272. Ullrich JT, McFeters GA, Temple KL (1972) J Bacteriol 110: 691–698

273. Lind DL, Daniel RM, Cowan DA, Morgan HW (1988) Enzyme Microb Technol 10: 180–186
274. Buonocore V, Sgambati O, De Rosa M, Esposito E, Gambacorta A (1980) J Appl Biochem 2: 390–397
275. De Rosa M, Gambacorta A, Nicolaus B, Buonocore V, Poerio E (1979) Biotech Lett 2: 29–34
276. Pisani FM, Rella R, Raia CA, Rozzo C, Nucci R, Gambacorta A, De Rosa M, Rossi M (1990) Eur J Biochem 187: 321–328
277. Bryant KJ (1986) MSc Thesis, University of Waikato, Hamilton, New Zealand
278. Cowan D, Daniel R, Morgan H (1985) TIBTech 3: 68–72
279. Fontana A (1988) Biophys Chem 29: 181–193
280. Heinen UJ, Heinen W (1972) Arch Mikrobiol 82: 1–23
281. Coolbear T, Daniel RM, Cowan DA, Morgan HW (1989) Annals NY Acad Sci 542: 279–281
282. Coolbear T, Eames CW, Casey Y, Morgan HW, Daniel RM (1991) J Appl Bacteriol 71: 252–264
283. Zamost BL, Brantley QI, Elm DD, Beck CM (1990) J Industrial Microbiol 5: 303–312
284. Cowan DA, Daniel RM (1982) Biochim Biophys Acta 705: 293–305
285. Khoo TC, Cowan DA, Daniel RM, Morgan HW (1984) Biochem J 221: 407–413
286. Cowan DA, Daniel RM (1982) Biotechnol Bioeng 24: 2053–2061
287. Cowan DA, Daniel RM, Morgan HW (1987) FEMS Microbiol Lett 43: 155–159
288. Cowan DA, Daniel RM, Morgan HW (1987) Int J Biochem 19: 741–743
289. Matsuzawa H, Hamaoki M, Ohta T (1983) Agric Biol Chem 47: 25–28
290. Matsuzawa H, Tokugawa K, Hamaoki M, Mizuguchi M, Taguchi H, Terada I, Kwon S-T, Ohta T (1988) Eur J Biochem 171: 441–447
291. Terada I, Kwon S-T, Migata Y, Matsuzawa H, Ohta T (1990) J Biol Chem 265: 6576–6581
292. Taguchi H, Hamaoki M, Matsuzawa H, Ohta T (1983) J Biochem 93: 7–13
293. Saravani G-A, Cowan DA, Daniel RM, Morgan HW (1989) Biochem J 262: 409–416
294. Janssen PH, Morgan HW, Daniel RM (1991) Appl Microbiol Biotechnol 34: 789–793
295. Cowan DA, Smolenski KA, Daniel RM, Morgan HW (1987) Biochem J 247: 121–133
296. Eggen R, Geerling A, Watts J, de Vos WM (1990) FEMS Microbiol Lett 71: 19–20
297. Lin X, Tang J (1990) J Biol Chem 265: 1490–1495
298. Fusek M, Lin X, Tang J (1990) J Biol Chem 265: 1496–1501
299. Cho H-Y, Tanizawa K, Tanaka H, Soda K (1987) Agric Biol Chem 51: 2793–2800
300. Cho H-Y, Tanizawa K, Tanaka H, Soda K (1988) J Biochem 103: 622–628
301. Minagawa E (1989) Jap J Dairy Food Sci 38: A259–A269
302. Hanner M, Redl B, Stoeffler G (1990) Biochim Biophys Acta 1033: 148–153
303. Freeze H, Brock TD (1970) J Bacteriol 101: 541–550
304. Danson MJ, Black SC, Woodland DK, Wood PA (1985) FEBS Lett 179: 120–124
305. Grossebüter W, Görisch H (1985) Syst Appl Microbiol. 6: 119–124
306. Sutherland KJ, Henneke CM, Towner P, Hough DW, Danson MJ (1990) Eur J Biochem 194: 839–844
307. Löhlein-Werhahn G, Goepfert P, Eggerer H (1988) Biol Chem Hoppe-Seyler 369: 109–113
308. Puchegger S, Redl B, Stoeffler G (1990) J Gen Micro 136: 1537–1542
309. Higa EH, Rameley RF (1973) J Bacteriol 114: 556–562
310. Nadal M, Mirambeau G, Forterre P, Reiter W-D, Duguet M (1986) Nature 321: 256–258
311. Kikuchi A, Asai K (1984) Nature 309: 677–681
312. Forterre P, Mirambeau G, Jaxel C, Nadal M, Duguet M (1985) EMBO Journal 4: 2123–2128
313. Kikuchi A, Shibata T, Nakasu S (1986) Syst Appl Microbiol 7: 72–78
314. Collin RG, Morgan HW, Musgrave DR, Daniel RM (1988) FEMS Microbiol Lett 55: 235–240
315. Slesarev AI, Kozyavkin SA (1990) J Biomol Struct Dyn 4: 935–942

316. Kovalsky K, Kozyavkin SA, Slesarev AI (1990) Nucl Acids Res 18: 2801–2806
317. Lehmacher A, Bisswanger H (1990) J Gen Microbiol 136: 679–686
318. Jones MD, Lowe DM, Borgford T, Fersht AR (1986) Biochemistry 25: 1887–1891
319. Sada E, Katoh S, Inoue T, Matsubara K, Shiozawa M, Akira T (1984) J Chem Eng Jpn 17: 642–646
320. Brand NJ, Fersht AR (1986) Gene 44: 139–142
321. Hara-Yokoyama M, Yokoyama S, Miyazawa T (1984) J Biochem 96: 1599–1607
322. Ankilova VN, Reshetnikova LS, Chernaya MM, Lavrik OI (1988) FEBS Lett. 227: 9–13
323. Garber MB, Yaremchuk AD, Tukalo MA, Egorova SP, Berthet-Colominas C, Leberman R (1990) J Mol Biol 213: 631–632
324. Garber MB, Yaremchuk AD, Tukalo MA, Egorova SP, Formenkova NP, Nikonov SV (1990) J Mol Biol 214: 819–820
325. Takahashi M, Yamaguchi E, Uchida T (1984) J Biol Chem 259: 10041–10047
326. Takahashi M, Uchida T (1986) J Biochem 100: 123–131
327. Engelke DR, Krikos A, Bruck ME, Ginsburg D (1990) Analyt Biochem 191: 396–400
328. Bergquist PL, Love DR, Croft JE, Streiff MB, Daniel RM, Morgan HW (1987) Biotechnology and Genetic Engineering Reviews 5: 199–244
329. Tanaka T, Kawano N, Oshima T (1981) J Biochem 89: 677–682
330. Taya M, Hinoki H, Kobayashi T (1985) Agric Biol Chem 49: 2513–2515
331. Honda H, Naito H, Taya M, Iijima S, Kobayashi T (1987) Appl Microbiol Biotechnol 25: 480–483
332. Honda H, Iijima S, Kobayashi T (1988) Appl Microbiol Biotechnol 28: 57–58
333. Soutschek-Bauer E, Staudenbauer WL (1987) Mol Gen Genet 208: 537–541
334. Love DR, Fisher R, Bergquist PL (1988) Mol Gen Genet 213: 84–92
335. Saul DJ, Williams LC, Love DR, Chamley LW, Bergquist PL (1989) Nucl Acids Res 17: 439
336. Saul DJ, Williams LC, Grayling RA, Chamley LW, Love DR, Bergquist PL (1990) Appl Environ Microbiol 56: 3117–3124
337. Luthi E, Love DR, McAnulty J, Wallace C, Caughey PA, Saul D, Bergquist PL (1990) Appl Environ Microbiol 56: 1017–1024
338. Neal TL (1987) MSc Thesis, University of Waikato, Hamilton, New Zealand
339. Luthi E, Jasmat NB, Bergquist PL (1990) Appl Environ Microbiol 56: 2677–2683
340. Luthi E, Jasmat NB, Bergquist PL (1990) Appl Microbiol Biotechnol 34: 214–219
341. Coleman RD, Yang S-S, McAlister MP (1987) J Bacteriol 169: 4302–4307
342. Koyama Y, Okamoto S, Furukawa K (1990) Appl Environ Microbiol 56: 2251–2254
343. Koyama Y, Furukawa K (1990) J Bacteriol 172: 3490–3495
344. Motoshima M, Azuma N, Kaminogawa S, Ono M, Minagawa E, Matsuzawa H, Ohta T, Yamauchi K (1990) Agric Biol Chem 54: 2385–2392

Biosynthesis of Storage Lipids in Plant Cell and Embryo Cultures

Nikolaus Weber[1], David C. Taylor[2] and Edward W. Underhill[2]

[1] Bundesanstalt für Getreide-, Kartoffel- und Fettforschung, Institut für Biochemie und Technologie der Fette, H. P. Kaufmann-Institut, Piusallee 68, D-4400 Münster, FRG

[2] National Research Council of Canada, Plant Biotechnology Institute, 110 Gymnasium Place, Saskatoon, Sask, Canada S7N OW9

The biosynthesis of storage lipids in plant cell and embryo cultures is discussed in the light of their significance in the breeding of agriculturally important oil seed crops. After a short introduction to the biosynthesis of storage lipids, i.e. triacylglycerols and wax esters, this review covers the occurrence and biosynthesis of storage lipids in plant cell and embryo cultures. Plant cells in culture generally contain low levels of both unusual fatty acids and triacylglycerols indicating that these cells are quite different from cells of oil storage tissues. There are a few exceptions to this rule which demonstrate that induction of genes involved in the expression of fatty acid modification and triacylglycerol assembly is possible in plant cell cultures. Such biosynthetically active plant cells may be of particular interest in future studies of storage lipid assembly. Both somatic and gametophytic embryos of oil plants exhibit high capacities for storage lipid biosynthesis and accumulation in vitro compared to cultured plant cells. Above all, the microspore-derived embryo system is recommended to both plant breeders and plant biochemists for the selection and multiplication of plants of superior quality.

Advances in Biochemical Engineering
Biotechnology, Vol. 45
Managing Editor: A. Fiechter
© Springer-Verlag Berlin Heidelberg 1992

1 Introduction

Some methods of unconventional plant breeding have recently gained importance for the development of improved crop plants. These methods include molecular biology and plant cell culture techniques which are eminently useful for isolating and multiplying genes from one organism and transferring them to another, e.g. a crop plant. Efforts are being directed, for instance, at transforming genetically traditional oil plants for improvement of agronomic performance and technical utility [1–6].

Vegetable oils consist mainly of mixtures of triacylglycerols differing in their fatty acid composition. Economically valuable fats and oils for oleochemical processes, however, require a fairly uniform composition of acyl moieties, often triacylglycerols enriched in a single type of acyl moiety. Progress in the genetic manipulation of oil plants may lead to triacylglycerols with such an altered composition of acyl moieties that is more useful for the oleochemical industry. To realize these ambitious biotechnological goals, it is necessary to understand the biosynthesis of fatty acids, the assembly of triacylglycerols in oil plants and the regulation of these processes, in detail.

In the present chapter, the benefits of the application of plant cell and embryo cultures in research on the biochemistry and physiology of triacylglycerol formation are discussed against the background of their significance in the breeding of oil plants.

2 Biosynthesis of Fatty Acids, Triacylglycerols, and Wax Esters in Plants

Triacylglycerols I (cf. Fig. 1) of traditional oil seed crops contain a limited number of different fatty acids; eight of these common fatty acids contribute more than 97% of the world production of edible vegetable oils [5]. Fatty acids such as palmitic and stearic acids are formed by fatty acid synthase in the plastids of plant cells.

It is generally accepted that a set of desaturases which is located in the plastids introduces double bonds into the acyl moieties bound as thioesters to acyl carrier protein or as oxoesters in monogalactosyl diacylglycerols. Both saturated and unsaturated fatty acids are translocated as acyl-CoAs from the plastids to the endoplasmic reticulum. Acyl chains of these activated fatty acids are elongated or incorporated into phosphatidylcholines and other polar lipids II (cf. Fig. 1), and partly modified, for example, by desaturation. Further modification of acyl moieties leads to hydroxy-, epoxy-, and cyclic fatty acids.

Triacylglycerols are assembled by the enzymes of the Kennedy pathway [6] which are located in the endoplasmic reticulum; the sequence of enzymatic reactions involved is shown in the following scheme [7].

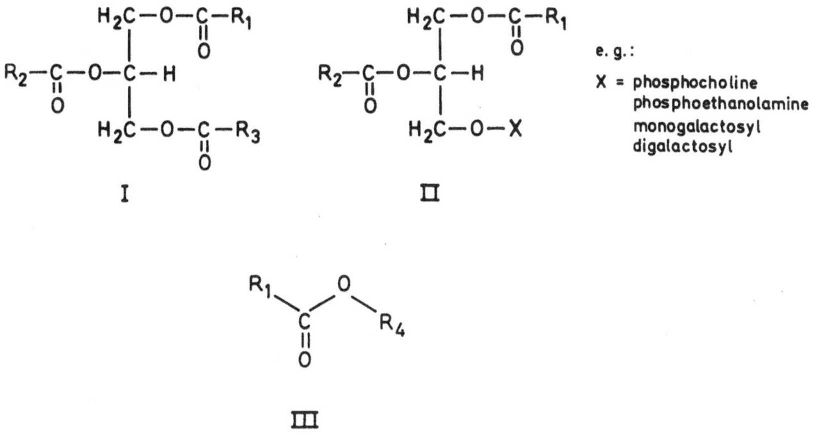

$$H_2C-O-\overset{O}{\underset{\|}{C}}-R_1$$
$$R_2-\overset{O}{\underset{\|}{C}}-O-\overset{|}{\underset{|}{C}}-H$$
$$H_2C-O-\overset{O}{\underset{\|}{C}}-R_3$$

I

$$H_2C-O-\overset{O}{\underset{\|}{C}}-R_1$$
$$R_2-\overset{O}{\underset{\|}{C}}-O-\overset{|}{\underset{|}{C}}-H$$
$$H_2C-O-X$$

II

e. g.:

X = phosphocholine
phosphoethanolamine
monogalactosyl
digalactosyl

$$R_1 \diagdown \diagup O \diagdown R_4$$

III

Fig. 1. Structure of triacylglycerols (I). i.e. the typical reserve lipids of oil seeds, and of ionic and non-ionic polar lipids (II), such as glycerophospholipids and glycerogalactolipids, which function as membrane lipids. Wax esters (III) are only formed in jojoba *(Simmondsia chinensis)* seeds as energy reserves. (R_1, R_2, R_3: various acyl moieties; R_4: alkoxy moiety; X: various polar head groups of ionic and non-ionic membrane lipids)

a. *sn*-glycerol-3-phosphate + acyl-CoA $\xrightarrow{\text{AT 1}}$ 1-acyl-2-lyso-*sn*-glycero-3-phosphates

b. 1-acyl-2-lyso-*sn*-glycero-3-phosphates + acyl-CoA $\xrightarrow{\text{AT 2}}$ 1,2-diacyl-*sn*-glycero-3-phosphates

c. 1,2-diacyl-*sn*-glycero-3-phosphates $\xrightarrow{-P_i}$ 1,2-diacyl-*sn*-glycerols

d. 1,2-diacyl-*sn*-glycerols + acyl-CoA $\xrightarrow{\text{AT 3}}$ triacyl-*sn*-glycerols

The acyl pattern of the final mixture of triacylglycerols is characteristic for each seed oil. The specific acyl chain targeting depends on both the selectivity of the various acyltransferases (AT 1, acyl-CoA: *sn*-glycerol-3-phosphate acyltransferase, E.C. 2.3.1.15; AT 2, acyl-CoA: 1-acyl-*sn*-glycero-3-phosphate acyltransferase, E.C. 2.3.1.51; AT 3, acyl-CoA: 1,2-diacyl-*sn*-glycerol acyltransferase, E.C. 2.3.1.20) of the Kennedy pathway and the compartmentation of triacylglycerol biosynthesis and/or acyl chain modification [6, 8].

Triacylglycerols occur predominantly in the seeds of oil plants, e.g. rapeseed *(Brassica napus)* and soybean *(Glycine max)*. In most higher plants, they are formed in the cotyledons of the embryo or the endosperm tissue during maturation and stored in subcellular organelles, i.e. oleosomes or oil bodies [9]. Triacylglycerols are cleaved by seed lipases which are activated during germination. The 'mobilized' fatty acids serve as sources of energy for the growth of the embryo during germination and provide biochemical intermediates for early metabolism, for example the biosynthesis of membrane lipids.

As an exception, wax esters III (cf. Fig. 1) in which long-chain fatty acids are esterified to long-chain alcohols are stored in seeds of jojoba *(Simmondsia chinensis)* as energy reserves [10]. Long-chain alcohols (fatty alcohols) are synthesized in plants by reduction of the corresponding acyl-CoAs. The final esterification of acyl-CoAs and fatty alcohols which is catalyzed by acyl-CoAs: fatty alcohol transacylase leads to wax esters [11].

3 Triacylglycerols in Plant Cell Cultures

Seed tissues of oil plants such as rape, soya, and coconut contain generally high proportions of fats and oils, i.e. triacylglycerols (I). Plant cells in culture, however, are rapidly growing organisms which synthesize predominantly ionic and nonionic polar lipids (II) for the assembly of new membranes.

3.1 Proportions of Triacylglycerols in Total Lipids

The contents of triacylglycerols in photosynthetic and non-photosynthetic plant cell cultures (Table 1) usually range from 5–10% of total lipids [12, 13]. Only a few examples of cultured plant cells with higher proportions of triacylglycerols are described in the literature (Table 1). Suspension cultures of carrot *(Daucus carota)* cells for instance, contain up to 30% of triacylglycerols in total lipids [14, 15]; even larger proportions are reported for cell cultures of *B. napus* [16] and *Papaver orientale* [17]. Cultured cells of anise *(Pimpinella anisum)* store extremely high proportions (around 100 mg g^{-1} dry weight) of these reserve lipids [18, 19]. It is of particular interest that callus cultures of cacao *(Theobroma cacao)* are able to double the content of triacylglycerols by incorporating exogenous stearic acid predominantly into these lipids [20]. However, the fatty acid composition of triacylglycerols of untreated callus and suspension cultures of cacao is similar to that in youngest embryos of cocoa beans [21, 22].

Table 1. Proportions of triacylglycerols in total lipids of various plant cell cultures

Plant species	Proportions (%, w/w) of triacylglycerols	Refs.
Ipomoea sp.	8	[23]
Tropaeolum majus	8	[24]
Brassica napus	8; 42 67	[16, 25]
Glycine max	5–13	[26–28]
Petroselinum crispum	8–13	[29]
Theobroma cacao	13; 17	[20, 30, 31]
Daucus carota	16; 30	[14, 15, 32]
Papaver orientale	5; 52	[17]
Pimpinella anisum	a	[18, 19]

a About 100 mg triacylglycerols per g dry weight, according to Refs. [18, 19]. This level is 10–15 times higher than levels of triacylglycerols found in most other plant cell cultures

The proportions of triacylglycerols in plant cell cultures are generally low. There are a few exceptions to this rule, as given in Table 1, indicating that induction of fatty acid biosynthesis and triacylglycerol assembly is possible in cultured plant cells. In some cases, accumulation of these lipids may be explained by induction of cell differentiation, e.g. embryogenesis (cf. Sect. 4) or by senescence-related lipid changes [33].

3.2 Subcellular Location of Triacylglycerols

Triacylglycerols of cultured plant cells are accumulated in oleosomes (Fig. 2) which are known to be cytoplasmic organelles of oil seeds; yet, other plant tissues also contain these lipid storing organelles. Oleosomes isolated from the cytosol of heterotrophic carrot suspension cells contain more than 97% of triacylglycerols [15], whereas cytosolic oleosomes from cultured cells of *Euphorbia tirucalli* consist mainly of steroids and terpenoids [34]. The subcellular distribution of oil bodies is different, however, in cultured photoautotrophic soya *(Glycine max)* cells. Biosynthesis and storage of triacylglycerols are located primarily in the chloroplasts of these cells [28].

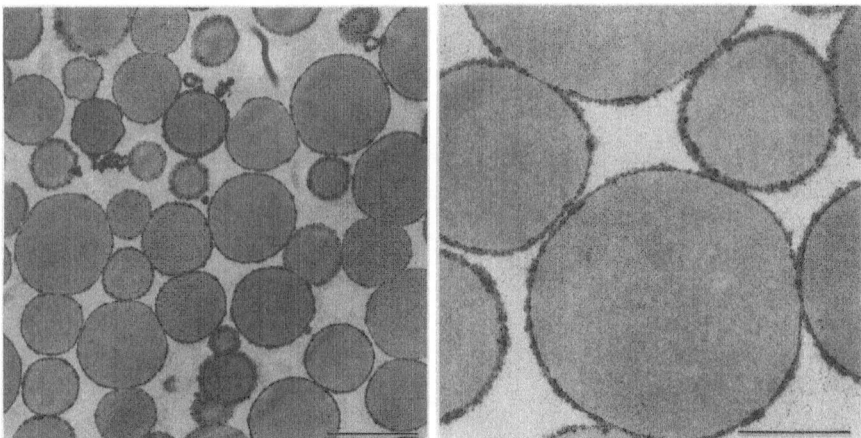

Fig. 2. Oleosomes (oil bodies) from heterotrophic cell suspension cultures of *D. carota* (electron micrographs; the bars indicate 1.0 μm, left, and 0.4 μm, right. From Kleinig et al. [15], with permission

3.3 Triacylglycerols Containing Unusual Fatty Acids

The constituent fatty acids of the various lipid classes occurring in cultured plant cells are usually similar to those found in the original plants. Exceptions to this rule are various seed tissues that contain triacylglycerols with certain unusual acyl

N. Weber, D. C. Taylor and E. W. Underhill

moieties, such as medium-chain, very long-chain, and cyclic acyl moieties [35]. It is known from a number of plant cell cultures that some of those unusual fatty acids which are constituent fatty acids of seed lipids are strongly excluded from the lipids of the cultured cells or, at least, occur only in traces in cell cultures established from these plants (Table 2).

For example, fatty acids containing hydroxy groups, e.g. ricinoleic acid ((Z)-9, 12-OH 18:1) in seeds of *Ricinus communis* [38], or certain positional and geometrical isomers of unsaturated fatty acids, e.g. petroselinic acid ((Z)-6 18:1) in seed lipids of *Pimpinella anisum* [17, 19], *Petroselinum crispum* [29] and *D. carota* [39], and eleostearic acid ((Z)-9, (E, E)-11, 13 18:3) in the seed lipids of *M. charantia* [36], are not detected in the lipids of cell cultures of these plants. Vaccenic acid ((Z)-11 18:1), however, is a constituent fatty acid of triacylglycerols and other lipids of *P. crispum* callus as well as various tissues of the intact plant, except seeds (Table 2) [29].

Epoxy fatty acids occur in seed oils of *Artemisia absinthium* and *Vernonia galamensis*, but not in callus cultures derived from these plants [40, 41]. In contrast,

Table 2. Composition of usual and unusual fatty acids in triacylglycerols and other lipids of seed tissues and plant cell cultures

Plant species	Tissue/ culture	Lipid fraction[a]	Composition (%) of fatty acids[b]								Ref.
			<16	16:0	18:0	18:1	18:2	18:3	20:1	22:1	
Momordica charantia	Cotyledons	TL		3	24	−	7	66[c]			[36]
	Callus	TL		52	12	16	10	10			
Cucumis melo	Cotyledons	TL		13	6	20	61	−			[36]
	Callus	TL	3	53	5	7	6	20	5[d]		
Tropaeolum majus	Seed	TL	Tr[e]	2	Tr	10	Tr	Tr	25	63	[24]
	Callus	TL	Tr	27	Tr	35	15	23	Tr	Tr	
Brassica napus	Seed	TAG	Tr	3	Tr	23	14	2	8	51	[16]
	Callus	TAG	Tr	18	5	29	29	12		2[f]	
Brassica napus cv. Rapora	Seed	NL		4	−	41	12	6	38[f]		[37]
	Suspension	NL		24	−	31	18	14	14[f]		
Pimpinella anisum	Seed	TL				70[g]	19				[17]
	Suspension	TL			22	10[g]	57				
Petroselinum crispum	Seed	TL	Tr	5	1	79[h]	14	Tr	1[f]		[29]
	Callus	TL	Tr	35	1	4[h]	44	7	9[f]		

[a] TL, total lipids; NL, neutral lipids; TAG, triacylglycerols. [b] Number of carbon atoms: number of double bonds; others add to 100%. [c] Eleostearic acid ((Z)-9, (E, E)-11, 13 18:3). [d] 20:0. [e] Tr, trace. [f] C > 18. [g] Isomeric 18:1 (%) in seeds: (Z)-6 = 54, (Z)-9 = 16; in suspension cells: (Z)-6 = <1; (Z)-9 = 10. [h] Isomeric 18:1 (%) in seeds: (Z)-6 = 86, (Z)-9 = 13, (Z)-11 = 1; in callus: (Z)-6 = 0; (Z)-9 = 60, (Z)-11 = 40

triacylglycerols and other glycerolipids containing cyclic fatty acids, e.g. cyclopropane and cyclopropene fatty acids in immature seeds of *Malva* spp., or very long-chain fatty acids (VLCFA), e.g. erucic acid (22:1) in *B. napus* seeds, occur in appreciable proportions in the lipids of cell cultures of these plants [37, 42]. Very long-chain fatty acids also have been detected in the lipids of cell cultures of *G. max* [42], *D. carota* [43], and *Chenopodium rubrum* [44]. Other authors, however, have found only small proportions of very long-chain fatty acids in the lipids of cell cultures of rape [16, 25, 45], soya [46], and carrot [46]. Cyclopentenyl fatty acids which are constituent fatty acids of the seeds of *Hydnocarpus anthelminthica* were only detected in trace amounts in total lipids of callus cultures of this plant [47]. In this context it is of interest that typical storage compounds of the seeds of *Euonymus europaea*, i.e. tri*acetyl*glycerols, are not synthesized by cell cultures of this plant [48].

In part, these different findings may be causally related to developmental processes in cultured plant cells leading to organogenesis or embryogenesis (cf. Sect. 4). Redifferentiation of such undifferentiated plant cells is known to be accompanied by alterations in lipid metabolism [49–51]. Developmental stimuli caused by plant hormones are known to be associated with phospholipid metabolism in cultured plant cells [52–54]. Little is known, however, about the hormone regulation of both assembly and accumulation of triacylglycerols in developing seeds. Abscisic acid which is involved in regulating storage protein accumulation [51, 55] may also be involved in accumulation of storage lipids [39] and senescence-related lipid changes [23].

3.4 Biosynthesis of Triacylglycerols in Plant Cell Cultures

Radioactively labeled precursors, e.g. acetate [56, 57] and fatty acids of various chain lengths and degrees of unsaturation [25, 57–60] as well as glycerol [14, 32], were used for studying biosynthesis and metabolism of glycerolipids, including triacylglycerols, in plant cell cultures.

Figure 3 shows the incoporation of different radioactive fatty acids into the triacylglycerols of rape and soya cells in suspension. Within seconds ("experimental zero time") after addition to the culture medium, appreciable amounts of exogenous fatty acids and labeled glycerolipids derived therefrom are detected inside the plant cells suggesting a fast translocation of fatty acids across the cell membranes. Unsaturated fatty acids are translocated across the membranes and incorporated into triacylglycerols and other lipids to a much higher extent than are saturated fatty acids [44]. It is worth noting that erucic acid was incorporated predominantly into the neutral lipid fraction of rape cells, though the lipids of these cells that have been derived from a low erucic rapeseed variety were obviously free of this very long-chain fatty acid [25].

Cell suspension cultures of cacao *(Th. cacao)* readily incorporated [1-^{14}C]acetate into both neutral and ionic glycerolipids [57]. After 24 h of incubation, around 18% and 45% of labeled acetate were incorporated into triacylglycerols and polar lipids, respectively. The incorporation of radioactive

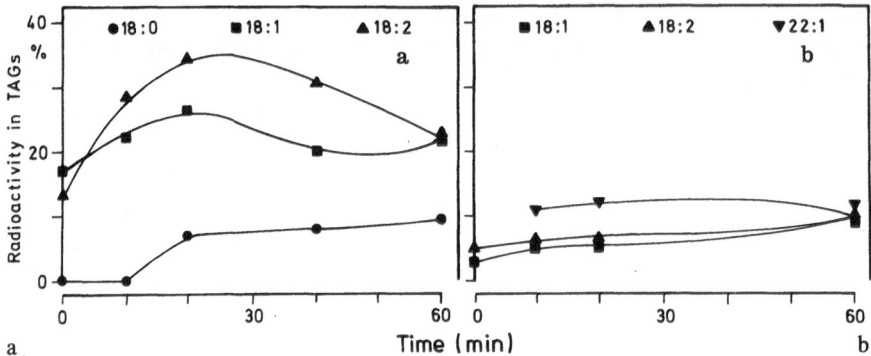

Fig. 3 a, b. Time course of the incorporation of various ^{14}C-labeled fatty acids into the triacylglycerols of soya cells (**a**) and rape cells (**b**) in suspension culture. After Stumpf and Weber [59] and Schneiders [25], respectively

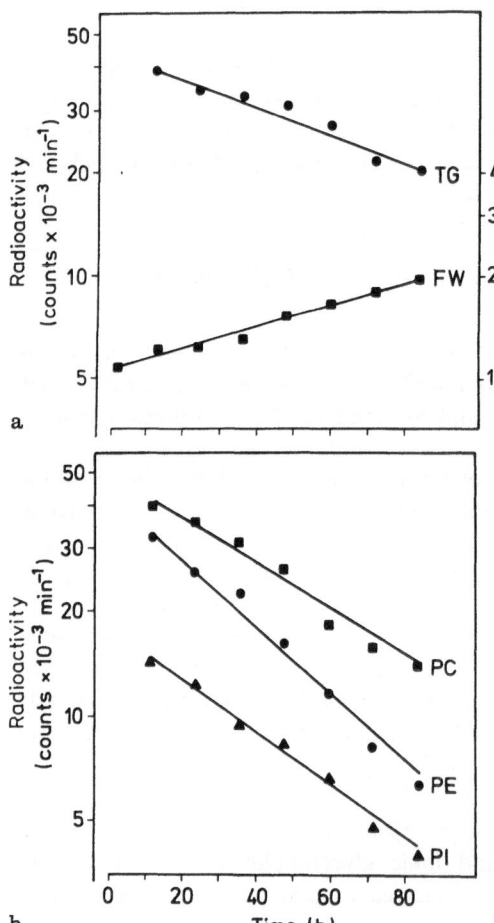

Fig. 4 a, b. Turnover of (**a**) tri-acylglycerols and (**b**) glycero-phospholipids in suspension cultures of *D. carota* cells in comparison with fresh weight (FW) of the cells as growth parameter (TG, triacylglycerols; PC, phosphatidylcholines; PE, phosphatidylethanolamines; PI phosphatidylinositols). From Kleinig and Kopp [14]

fatty acids, such as lauric, palmitic, stearic, oleic, and linoleic acids, into triacyl-glycerols and other lipid classes of cultured cacao cells was similar to that observed for other plant cell cultures [57, 58]. The tendency of cultured cacao cells to desaturate oleic acid and to incorporate exogenous fatty acids predominantly into phospholipids clearly showed that these cells were not suitable for the formation of triacylglycerols similar to cocoa butter [58].

The turnover of radioactive triacylglycerols was studied in heterotrophic carrot cells in suspension culture (Fig. 4) in order to determine their involvement in cellular lipid metabolism [14, 15]. It is evident from the data given in Fig. 4 that the half-life of triacylglycerols in these cells is about 70 h, whereas that of polar lipids is much shorter. On the other hand, generation time of the cells is higher than the half-life time of triacylglycerols demonstrating a considerable metabolism of these 'storage' lipids. Similar results are reported by Martin et al. [28] in photoautotrophic soya cells and by De Silva and Fowler [61] in heterotrophic sycamore *(Acer pseudoplatanus)* cells suggesting a rather temporary storage of triacylglycerols for later membrane biosynthesis than for storing energy. Surprisingly, the chloroplasts of photoautotrophic soya cells [28] are the primary location of acyl-CoA: 1,2-diacylglycerol acyltransferase (cf. Scheme in Sect. 2) which is known to be a typical enzyme of the endoplasmic reticulum of oil seeds.

Cell suspension cultures of soya were incubated with mixtures of positional and geometrical isomers of [1-^{14}C]octadecenoic acids in order to determine enzymatic specificities for the incorporation of individual isomeric fatty acids into the glycerolipids, for instance triacylglycerols [60]. It is obvious from the data given in Table 3 that $\Delta 9$ isomers of both mixed substrates, *(Z)*- and *(E)*-octadecenoic acids, are preferentially incorporated into the *sn*-2 position of the glycerol backbone, whereas a much lower percentage of these acyl moieties are esterified to the

Table 3. Incorporation of positional and geometrical isomers of [1-^{14}C]octadecenoic acids into triacylglycerols of cultured soya cells [60]

Lipid	Distribution (%) of radioactivity in positional isomers								
	$\Delta 8$	$\Delta 9$	$\Delta 10$	$\Delta 11$	$\Delta 12$	$\Delta 13$	$\Delta 14$	$\Delta 15$	$\Delta 16$[a]
(Z)-[1-^{14}C]octadecenoic acids (mixed substrates)[c]	3	21[b]	9	9	33	7	5	13	—
Triacylglycerols									
total acyl moieties	4	22[b]	9	8	34	8	5	10	
sn-2 acyl moieties	—	56[b]	6	8	20	4	2	3	
(E)-[1-^{14}C]octadecenoic acids (mixed substrates)[c]	5	13[b]	12	12	19	11	11	10	5
Triacylglycerols									
total acyl moieties	5	13[b]	11	13	17	13	11	12	5
sn-2 acyl moieties	2	31[b]	16	8	19	9	6	5	3

[a] Position of double bonds. [b] *(Z)*-9 18:1, oleic acid; *(E)*-9 18:1, elaidic acid. [c] Distribution of the various isomers in the mixed substrates for comparison

sn-1,3 positions. Apparently, acyl-CoA : 1-acyl-*sn*-glycero-3-phosphate acyltransferase of the Kennedy pathway (cf. Scheme) specifically transfers the two isomeric Δ9-octadecenoic acids, i.e. oleic acid (*(Z)*-9 18 : 1) and elaidic acid (*(E)*-9 18 : 1), to the *sn*-2 position.

In addition to the precursors mentioned above, 1-*O*-alkyl-*sn*-glycerols and 2-*O*-alkyl-*sn*-glycerols have been used as precursors for studying the metabolism of both neutral and ionic ether glycerolipids in plant cell cultures [62–64]. The cells of higher plants do not contain ether glycerolipids. They incorporate, however, exogeneous alkylglycerols into alkyldiacylglycerols, i.e. ether analogs of triacylglycerols, and other ether lipids. The enzymatic reactions involved in the metabolism of alkylglycerols are highly stereospecific; for example, starting from *rac*-1(3)-*O*-alkylglycerols 1-*O*-alkyl-2,3-diacyl-*sn*-glycerols are formed in rape suspension cells. The alkyl moieties of ether glycerolipids are metabolically quite stable, as compared to acyl moieties of ester glycerolipids. Therefore, alkyldiacylglycerols generated from exogeneous alkylglycerols in plant cell cultures may be useful markers in studies on the biosynthesis and turnover of storage lipids in plant cells. The usefulness of an equimolar mixture of homologous saturated 1-*O*-alkylglycerols for the determination of chain length specificity of enzymes which are involved in the formation of 1-*O*-alkyl-2-acylglycerols and 1-*O*-alkyl-2,3-diacylglycerols in cultured rape cells is demonstrated in Fig. 5.

Fig. 5a shows that 1-*O*-octadecyl species of the mixed 1-*O*-alkylglycerol substrate were preferentially incorporated into 1-*O*-alkyl-2-acylglycerols, whereas species containing alkyl moities with 14, 16, and 20 carbon atoms were almost equally distributed. 1-*O*-Alkyl-2,3-diacylglycerols which had been formed in rape cells from the equimolar mixture of 1-*O*-alkylglycerols preferentially contained alkyl moities with 16 and 18 carbon atoms and low proportions of species containing alkyl moieties with 14 and 20 carbon atoms (Fig. 5b).

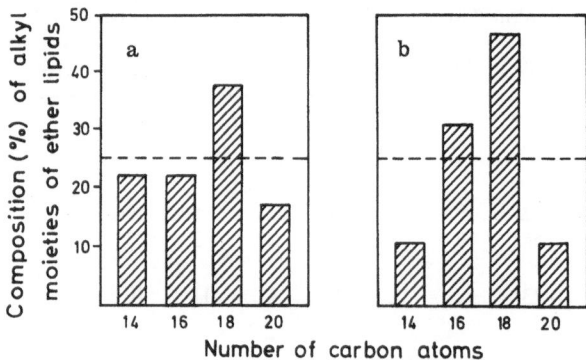

Fig. 5 a, b. Incorporation of homologous saturated 1-*O*-[1-^{14}C]alkylglycerols from an equimolar mixture into (**a**) 1-*O*-alkyl-2-acyl-*sn*-glycerols and (**b**) 1-*O*-alkyl-2,3-diacyl-*sn*-glycerols of photomixotrophic cell suspension cultures of *B. napus*. (The *dotted line* represents the percentage composition of each substrate in the equimolar mixture.) From Weber and Mangold [62]

It is obvious from the low levels of triacylglycerols and the relatively high proportions of ionic and other polar lipids in most cultured plant cells that these cells are quite different from cells of oil storage tissues. The ultrastructure of these cells shows that they are rich in plasma membranes, as is characteristic of undifferentiated cells. Most likely, the induction of genes involved in early differentiation of the cells is necessary in order to express the various enzymatic reactions of fatty acid modification and triacylglycerol assembly, as is evident in undifferentiated anise cells [18, 19] and various other embryogenic plant cell cultures [14, 49–51] (cf. Sect. 4). Such biosynthetically active cells may be of particular interest in future studies of storage lipid assembly, for example in genetically modified cells of oil plants.

4 Triacylglycerols and Wax Esters in Plant Embryo Cultures

Both somatic and microspore-derived embryos of oil plants are valuable tools in applied lipid biochemistry. It is well known from the literature that cultured embryos of various oil plants accumulate triacylglycerols during maturation and that they synthesize unusual fatty acids found in the seeds. The capacity of cultured embryos and embryogenic cells of oil plants as model systems is, therefore, reviewed with respect to their usefulness in studying the mechanisms of genetic control of biosynthesis and assembly of triacylglycerols in oil seeds.

4.1 Morphological and Physiological Aspects of Embryogenesis in Oil Seeds

In plants, triacylglycerols are stored in the fruit tissues (mesocarp, pericarp), e.g. the mesocarp of avocado, or in the seed tissues, e.g. the endosperm of poppy seeds and the cotyledons of rape seeds, or in both fruit and seed tissues, e.g. mesocarp and endosperm ('kernel') of oilpalm [7]. In accordance with the location of their oil storage tissues, oil seeds can be divided morphologically into two types, i.e. endospermic and non-endospermic (cotyledonary) oil seeds (Fig. 6).

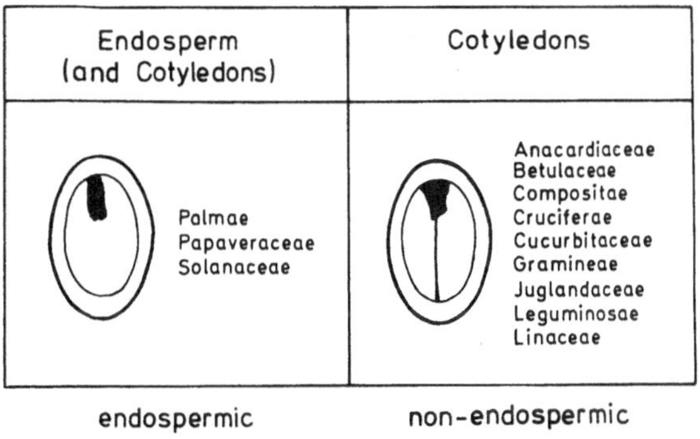

Endosperm (and Cotyledons)	Cotyledons
Palmae Papaveraceae Solanaceae	Anacardiaceae Betulaceae Compositae Cruciferae Cucurbitaceae Gramineae Juglandaceae Leguminosae Linaceae
endospermic	non-endospermic

Fig. 6. Main types of oil seeds, i.e. left, endospermic and right, non-endospermic (cotyledonary) oil seeds. After Thies [65]

Endospermic oil seeds such as castor bean consist of oil-rich endosperm and embryo which remain separate tissues from fertilization to maturity. In non-endospermic oil seeds such as rapeseed and soybean, the endosperm is increasingly displaced by the cotyledons of the developing embryo during seed maturation. Biosynthesis and accumulation of triacylglycerols are found to be primarily located in the endosperm tissue of the former and the embryonic cotyledons of the latter seeds.

Asexual or somatic embryogenesis, a natural phenomenon in many species [66] can be defined as the process of embryo initiation and development from cells that are not the direct products of gametic fusion [67]. Gametophytic embryos are of two types: androgenic and gynogenic. Both arise from post-meiotic cells, the former from microspores [68] and the latter from megaspores and their derivatives [69]. Gynogenic embryos have not been studied extensively at the physiological or biochemical level, perhaps because of the inherent difficulties in their induction and isolation [70]. However, androgenic embryos, and in particular, microspore-derived embryos, have been obtained from a number of species via the culture of immature anthers or isolated microspores.

Both somatic (asexual) and gametophytic embryogenesis (Fig. 7) in vitro represent developmental processes which parallel in many respects those of zygotic embryogenesis in seeds subsequent to fertilization. As a consequence, the cellular structures of developing embryos derived from maturing seeds are similar to those of somatic and gametophytic embryos growing in culture. The various morpho-genetic stages observed during embryogenesis of zygotic embryos are closely related to those found during the development of both somatic and gametophytic embryos, e.g. microspore-derived embryos of *B. napus* (Fig. 8).

Maturation of zygotic embryos of oil seeds is generally characterized by three different phases:

a. high cell proliferation in seed tissues after fertilization,
b. induction of biosynthesis and accumulation of triacylglycerols (and other energy reserves, e.g. proteins), and finally,
c. completion of biosynthetic activities, desiccation, and seed dormancy.

Origin of Embryo	Seed	Plant Cell Culture[a]	Microspore/ Macrospore
	↓	↓	↓
Type of Embryo	Zygotic	Somatic	Gametophytic (androgenic/ gynogenic)
Genotype of Embryo	heterozygotic, diploid, heteroploid	homo- or hetero- zygotic, diploid (haploid)	homozygotic, haploid (diploid)

[a] Derived from various meristematic tissues

Fig. 7. Classification of plant embryos according to their origin

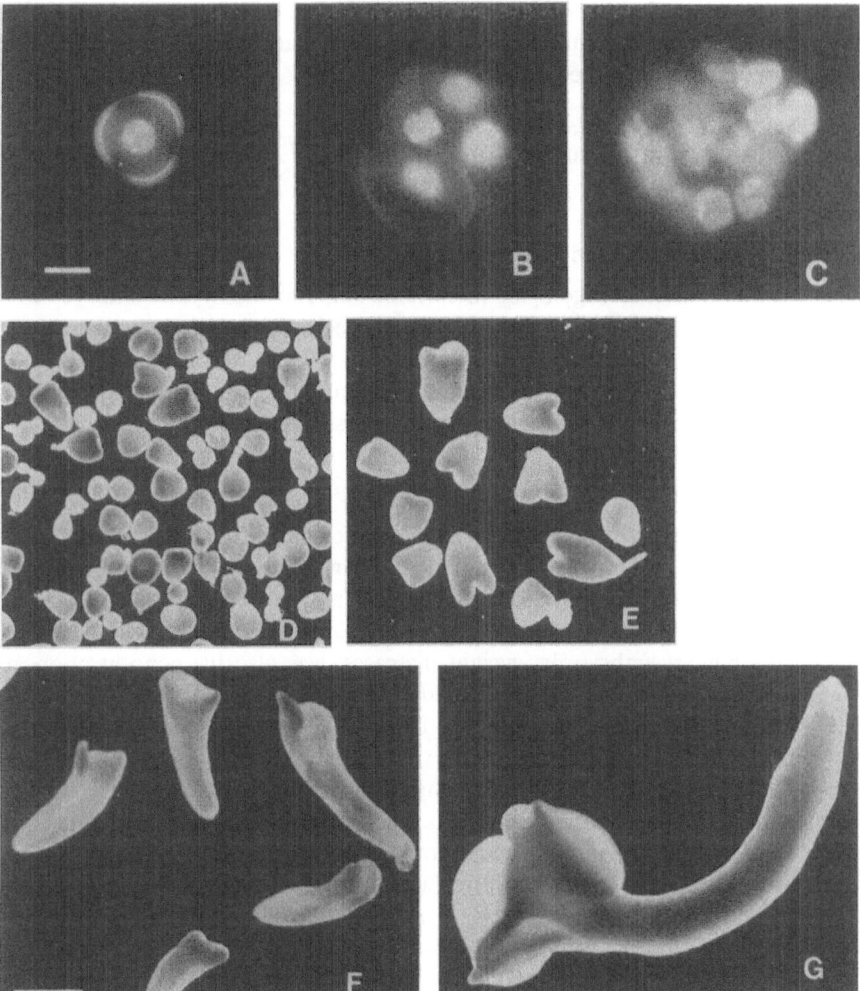

Fig. 8 A–G. Morphogenetic stages of *B. napus* microspore-derived embryos. A–C are early stage material fixed briefly (glacial acetic acid/95% ethanol: 3/1), resuspended in water and stained for DNA with 2.5 μg ml^{-1} 4,6-diamidino-2-phenylindole. Photos A–C were taken under EPI-illumination and are on the same scale (*bar* in A = 10 μm) **A**: freshly isolated uninucleate microspore; **B**: taken from a 5-d culture; **C**: taken at 10 d in culture (globular stage) with 60–70 nuclei. Photos D–G were taken on a Zeiss IM inverted microscope with phase ring and are on the same scale (*bar* in F = 500 μm). **D**: globular embryos; **E**: heart-shaped embryos; **F**: early torpedo-shaped embryos; **G**. cotyledonary stage of embryo development. Photographs courtesy of Dr. Larry A. Holbrook, P.B.I., NRC Canada

In cultured somatic and gametophytic embryos of oil plants similar developmental stages are observed, particularly with respect to the formation and accumulation of triacylglycerols and other reserve compounds, as discussed below. However, in cultured embryos, dessication and dormancy phases are not typically observed;

rather, prevention of precocious germination or a switch over to germinative type metabolism are usually controlled by adjusting the relative proportions of growth regulator compounds in the culture medium.

4.2 Occurrence and Composition of Storage Lipids in Plant Embryo Cultures

4.2.1 Cultured Somatic and Zygotic Embryos

The accumulation of storage lipids in cultured somatic and zygotic embryos will be discussed for a number of important oilseed species. The primary focus of work with these embryos has been the production of "seed-type" lipids in culture, and, in the case of palm, selection of desirable clones. Particular emphasis will be placed on the deposition of storage lipids containing "unusual fatty acids" (cf. Sect. 3.3).

4.2.1.1 Brassica spp.

Recently, it was reported that zygotic and somatic embryos of both *B. napus* and *B. campestris*, cultured in vitro, accumulated storage lipids [71]. This accumulation could be influenced by culture conditions. During in vitro culture of *B. napus* zygotic embryos, the sucrose concentration (10%) in the culture medium which yielded the highest embryo dry weight increases, was also the concentration at which the highest amount of storage lipid (41.2 mg g^{-1} dry weight) accumulated. The pattern of storage lipid accumulated by *B. napus* somatic embryos in culture, was somewhat different, reaching a maximum of 21.2 mg g^{-1} dry weight at a sucrose concentration of 20%. The ranges of sucrose concentrations eliciting storage lipid accumulation are similar to those reported for cultured embryos of jojoba (c.f. Sect. 4.2.1.4). Avjioglu and Knox [71] stated that the amounts of storage lipid observed in their embryo cultures are higher than the triacylglycerol level present in the mature seed (13.1 mg g^{-1} dry weight). However, it would seem that the latter value, as cited, is at least 30-fold lower than values reported elsewhere (300–450 mg g^{-1} dry weight) for mature seed of *B. napus* [51, 72, 73]. Thus, the capacity for cultured zygotic and somatic embryos of *B. napus* to produce storage lipids in culture at rates approaching those observed in zygotic embryos in vivo, remains questionable.

The acyl composition of somatic embryos from *B. campestris* cultured in 10% sucrose, was similar to that found in immature zygotic embryos in vivo and in mature seed. Of note, somatic embryos also accumulated the very long chain "seed-specific", eicosenoic and erucic acids [71]. It has also been reported that zygotic embryos of *B. juncea*, cultured in vitro accumulate lipids with a fatty acyl composition similar to the embryo in vivo [74].

4.2.1.2 Borage (Borago officinalis L.)

In vivo, 10- and 24-day zygotic embryos of borage possess a total fatty acid content of about 18% and 32% dry weight, respectively [75]. 10-Day embryos cultured in vitro on liquid media containing 6% sucrose for 14 d, reached a peak of total

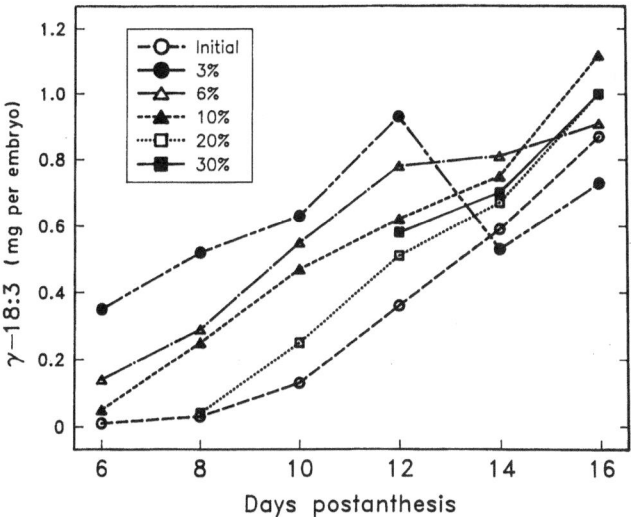

Fig. 9. Accumulation of γ-linoleic acid (γ-18:3) by zygotic embryos of *Borago officinalis* cultured in basal medium supplemented with various sucrose concentrations (%, w/v, indicated in inset) from 6 to 16 d post anthesis. Data from liquid and semisolid media are combined. After Whipkey et al. [76]

fatty acid content at 41% of the dry weight. The concentration of γ-linolenic acid (γ-18:3) in vivo was 11.2% of the total fatty acids at 10 and 24 d post-anthesis, but γ-18:3 content varied from 8 to 18% when 10-d old embryos were cultured for 14 d, depending on the media state (liquid or solid) and sucrose concentration. The highest concentration of γ-18:3 in vitro was produced on semi-solid media at 3% sucrose (Fig. 9), and γ-18:3 accumulation decreased with increasing sucrose concentration [76].

Cotyledonary zygotic embryos cultured in basal medium with 4.5 μM 2,4-D plus 10% coconut water were found to initiate somatic embryos in 8 weeks [77]. Somatic embryos originated directly from the surface of cotyledons, and indirectly from a mass of proembryonic callus. A non-browning embryogenic clone cultured in 2,4-D continued to produce yellow embryogenic callus and partially differentiated somatic embryos. When transferred to 2,4-D-free medium and cultured for 30 d, some cotyledonary embryos developed. However, the frequency of cotyledonary embryos was relatively low compared to the other morphological tissue types (green leafy growth, translucent globular structures, white nodular structures, white callus) arising from the various culture conditions. Of all the tissues examined, cotyledonary embryos contained the highest proportion (about 23%) of total fatty acids on a dry weight basis. While all tissues examined contained γ-18:3, the cotyledonary embryos were highest at about 20% [77]. Thus, selection of non-browning clones for higher rates of conversion to cotyledonary embryos in vitro could provide an alternative system to produce γ-18:3.

4.2.1.3 Cacao (Theobroma cacao L.)

Early studies by Janick's group indicated that both zygotic and asexual embryos cultured in vitro could be shifted to a phase of storage lipid accumulation by addition of high concentrations of sucrose to the basal medium [78, 79]. The protocol of choice [78] was to place embryos (equivalent to 100–120 d post-pollination) in liquid basal medium containing 3% sucrose for 10 d. Then, cultures were shifted to basal medium containing 9, 15, and 21% sucrose for 2 d each, with a final 30 d culture period in the presence of 27% sucrose. The high sucrose protocol employed at day 10 induced the suppression of growth and the simultaneous accumulation of lipids [79] as well as anthocyanins, alkaloids and proteins [67]. Cultured zygotic and asexual (somatic) embryos responded to this protocol in the same manner. Fatty acid and triacylglycerol accumulation in both asexual and zygotic embryos cultured in vitro at 26 °C followed a similar pattern of development to that observed for zygotic embryos in vivo [80–82] (Table 4). Fatty acid quality equivalent to cocoa butter was not obtained until total fatty acids in embryos approached 50% on a dry weight basis, and the latter criterion was used to gauge embryo maturity in culture. The protocol for development of

Table 4. Embryo growth and lipid production of asexual (somatic) embryos of cacao *(Th. cacao)* cultured in low and high sucrose medium compared with zygotic embryos in vivo. Adapted from Refs. [80–82]

Variables[a]		Lipid production in		
		Asexual embryo		Zygotic embryo
		Sucrose protocol (% Sucrose in culture medium, w/v)		In vivo
		3%	3–27%	
Embryo growth				
FW	(g)	2.5	1.0	1.5
DW	(g)	1.0	0.4	1.0
FW/DW	(%)	8.0	40.0	68.0
Lipid production				
TL/DW	(%)	4.1	26.0	48.0
FA/DW	(%)	1.2	24.0	45.0
FA/TL	(%)	29.0	92.0	93.0
FA composition (mol%)				
Palmitic	16:0	34	26	29
Stearic	18:0	11	29	33
Oleic	18:1	5	34	34
P + S + O[b]		50	89	96
Linoleic	18:2	40	9	4
Linolenic	18:3	11	>1	>1

[a] DW, dry weight; FA, fatty acids; FW, fresh weight; TL, total lipids. [b] P + S + O = Sum of palmitic, stearic, and oleic acids which are the constituent fatty acids of cocoa butter

Table 5. Accumulation of triacylglycerols[a] in embryos of cacao *(Th. cacao)* and composition of fatty acids: A comparison of asexual (somatic) and zygotic embryos both cultured in sucrose containing medium[b]. Adapted from Ref. [67]

Source of embryo	Accumulation of fatty acids (% of DW)	P + S + O[c] (Mol%)
Zygotic in vivo	52.0	94.0
	Best individual embryos	
Zygotic in vitro	43.4	93.9
Asexual in vitro	39.2	92.9
	Best experimental mean	
Zygotic in vitro	32.4	90.3
Asexual in vitro	25.3	89.6
	Mean (12 experiments)	
Asexual in vitro	16.9	85.3

[a] Expressed as accumulation of fatty acids. [b] Sucrose protocol described in detail in Ref. [78]. [c] cf. Table 4

asexual embryos was characterized by large variability in embryo dry weight and fatty acid production on a dry weight basis (Table 5). Futhermore, the embryos did not reach maturity. Tests of various culture conditions including nutrient modifications [83] and physical parameters [84] failed to indicate the limiting factors. The in vitro production of cocoa butter will require a greater understanding of the regulation of lipid biosynthesis in embryo culture systems and will ultimately depend on the suitability of the metabolic product and the economics of the system [67].

4.2.1.4 Jojoba *(Simmondsia chinensis* Link*)*

Jojoba, a xerophytic shrub native to arid regions of south western North America, stores lipid in the form of wax esters of long chain $(C_{16}-C_{24})$ fatty acids and $(C_{18}-C_{24})$ alcohols [85]. In vivo, zygotic embryos of jojoba accumulate these liquid wax esters from early development (10 mg dry weight) to maturity (800 mg dry weight) with relatively constant proportions of 36–46 carbon wax esters. Thus, jojoba liquid wax of good quality can be obtained by harvesting zygotic embryos at any stage of maturation. At maturity, wax esters comprise >50% of seed dry weight [86].

Immature zygotic embryos (1–2 mg dry weight) cultured on a semisolid basal medium supplemented with 15% sucrose for 10 weeks, reached a dry weight of 65 mg and produced 50% wax esters, an efficiency twice that of in vivo zygotic embryos of a comparable size. Somatic embryos (0.3 mg dry weight) cultured on basal medium containing 9% sucrose for 12 weeks, averaged about 95 mg dry weight with 20% wax esters, representing an efficiency of about 75% compared to zygotic embryos of the same size in vivo [86] (Table 6). The effect of added sucrose as a stimulant of storage lipid accumulation is similar to that observed in cultured embryos of *Theobroma cacao*, discussed above. The proportion of wax

Table 6. Wax esters in jojoba *(Simmondsia chinensis)*: A comparison of somatic embryos grown in vitro with zygotic embryos grown in vivo and in vitro. After Ref. [86]

Source of embryos	Growth period	DW	Wax ester content	Wax ester accumulation	In vitro efficiency[a] (% in vivo)	
					Immature basis	Mature basis
	(weeks)	(mg)	(mg)	(% DW)		
Zygotic in vivo	≈ 26	≈ 500	≈ 250	≈ 50	–	–
	In vitro mean					
Zygotic	10	65	32	50	198	13
Somatic	12	93	19	20	74	8
	Best embryos in vitro					
Zygotic	10	141	88	62	198	35
Somatic	12	214	65	30	84	26

[a] In vitro wax ester production by comparable-sized embryos based on dry weight (DW)

esters accumulated by both zygotic and somatic embryos cultured in vitro, was similar to that of jojoba wax esters produced in vivo. With efforts to optimize protocols for somatic embryo development and culture conditions, the prospects will improve for liquid wax ester formation in vitro.

4.2.1.5 Soybean *(Glycine max* L.*)*

In mature zygotic embryos of soybean, linoleic acid *((Z)*, *(Z)*-9, 12 18:2) is the predominant fatty acid [87]. Recent studies by Shoemaker and Hammond [88] have shown that somatic embryos, initiated from immature cotyledons of zygotic embryos, also possessed high levels (>55%) of 18:2. The relative percentages of fatty acids in lipids of somatic embryos from a number of cultivars indicated that, as in zygotic seed embryos of soybean, variation in the percentage of palmitic, oleic and linoleic acid were affected by length of maturation (30 vs 45 d in "maturation medium") but were genotype-dependent.

4.2.1.6 Oil Palm *(Elaeis guineensis* Jacq.*)*

Since the oil palm is a monocotyledon, it grows only from the central meristem and callus production is difficult [31]. However, in 1972, Rabechault et al. [89] reported the development of shoots from palm crown tissue cultures and Jones [90] obtained embryoids from callus of root and leaf base which were subsequently regenerated into plantlets. Somatic embryogenesis from leaf tissue of seedlings cultivated in vitro was first reported in 1976 [91]. The current methods available for oil palm propagation have been reviewed elsewhere, recently [92]. Embryoids formed on both primary and secondary calluses and isolated embryoids can be induced to proliferate to form a continuously propagating morphogenetic tissue with the capacity to produce plants (clones) over many subcultures [93]. Histological examination of somatic embryos arising from leaf explants revealed a high content of storage lipid early on in embryogenesis, i.e. when the cambium-like zone grew protuberances in which there was nothing from the anatomical point of view to indicate that they would subsequently become embryoids. Thus, the accumulation of storage lipids appears to be an early indicator that tissues cultured in vitro are developing into true somatic embryos [94].

The oils produced by a range of different clones grown in Malaysia have been analyzed [95]. The clones were derived from early tissue culture experiments and were not from selected palms. While there was a wide variation in oil composition among the clones, there was a high within-clone uniformity, indicating a strong genetic control of oil composition. Thus, selection of clones for oil quality should be possible.

4.2.1.7 Other Species

Upon induction of somatic embryogenesis in poppy, *Papaver* spp., storage lipid accumulation was observed [49, 50]. As a measure of metabolic changes during the process of redifferentiation, the lipid content was monitored. While changes in phospholipid content were less distinct, a strong and characteristic accumulation of storage lipid was correlated with globular body formation [49]. In somatic embryos of *Papaver orientale*, triacylglycerol levels peaked at about 30 μmol g^{-1} dry weight at 30 d after induction (days of regeneration) and subsequently fell to a pre-induction level of 5 μmol g^{-1} dry weight. The latter event was associated with a switch to germinative metabolism and plantlet formation.

Cell cultures of anise, *Pimpinella anisum*, accumulated lipids and were easily induced to form somatic embryoids in high yield in liquid culture medium [96, 97]. Electron micrographs indicated a large number of lipid bodies per cell. Seed embryos contained about 20% triacylglycerols on a dry weight basis, while somatic embryos contained about 5%, mainly in the cotyledonary primordia. In the storage lipid of anise seeds, the major fatty acids were *(Z)*-6 18:1 (petroselinic acid) (54%), *(Z)*, *(Z)*-9, 12 18:2 (19%), *(Z)*-9 18:1 (16%) and 16:0 (4%). Cell cultures of anise have a very low petroselinic acid content (<1%), but upon undergoing somatic embryogenesis, the proportion of this unusual fatty acid typical of anise seeds increased dramatically to a level of about 18% [19]. Work continues on the in vitro synthesis of triacylglycerols containing petroselinic acid in somatic embryoids derived from a number of cell culture lines.

In triacylglycerols of carrot, *Daucus carota*, seeds petroselinic acid is present in large amounts (>70%) [98], but, as discussed above, cell suspension cultures of this species do not contain this fatty acid. However, recent studies [39] have shown that petroselinic acid was produced to a limited extent in somatic embryos, induced from carrot callus cultures, and again it has been esterified exclusively to triacylglycerols. Petroselinic acid first appeared at the heart stage and reached a maximum at 1.4% in torpedo-shaped embryoids. Different culture conditions such as addition of abscisic acid, sorbitol and these in combination, gave the maximum level of total lipids and triacylglycerols (17 mg g^{-1} fresh weight) in embryoids. Since the levels of petroselinic acid were low in late-stage embryoids, but absent in undifferentiated tissues, it was concluded that there must be weak, but distinct, embryo-specific gene expression governing the biosynthesis of this fatty acid, and its deposition in triacylglycerols.

4.2.2 Cultured Gametophytic Embryos

4.2.2.1 Gynogenic Embryos

Little has been reported on the occurrence of storage lipids in cultured gynogenic embryos. However, fertilized tobacco (*Nicotiana tabacum* L.) ovaries in culture were reported to accumulate fatty acids such that the level at embryo maturity in vitro was about 25% of zygotic embryos allowed to mature in vivo on the mother plant [99]. The acyl composition of total lipids and individual lipid classes was similar in ovaries cultured at 28 °C and seeds grown in the greenhouse at the same temperature (Table 7).

Table 7. Fatty acid composition of total lipids and individual lipid classes in greenhouse-grown seeds and cultured ovaries of *Nicotiana tabacum* cv. BY-4. After Matsuzaki et al. [99]

Seeds	Lipids[a]	Fatty acid contents ($\mu g\ mg^{-1}$ lipid)	Fatty acid composition (%, w/w)				
			16:0	18:0	18:1	18:2	18:3
Greenhouse at 28 °C	TL	603	9.7	3.0	13.6	72.7	1.0
	TAG	553	9.6	3.3	14.1	72.1	0.9
	DAG	7	16.5	7.5	20.0	56.0	Tr
	PL	8	22.0	5.8	25.2	47.0	Tr
Ovary culture at 28 °C	TL	544	7.8	2.8	17.8	70.8	0.8
	TAG	471	7.7	2.7	17.5	71.2	0.9
	DAG	7	13.2	5.0	24.0	57.8	Tr
	PL	11	22.5	7.1	27.0	42.1	1.3

[a] DAG, diacylglycerols; PL, polar lipids; TAG, triacylglycerols; TL, total lipids

4.2.2.2 Microspore-Derived (Androgenic) Embryos of Brassica spp.

Research with microspore-derived (MD) embryos, has focused on exploiting this system for breeding and selection and biochemical studies. Progress in these areas will be reviewed, with a focus on the biochemistry that has emerged from studies of these haploid embryos.

Production of MD embryos from anther culture of rape *(B. napus)* was first reported in the mid-to late 1970s [100–102]. Early studies using a small number of anther-derived embryos of *B. campestris* revealed that erucic acid (*(Z)*-13 22:1) was present and it was suggested that such gametophytic embryos might be utilized to study storage lipid biosynthesis [103]. High frequencies of embryogenesis are now available in *Brassica* spp. [68, 104, 105]. The relative ease with which high-quality haploid embryos of *B. napus* can be generated makes this system an excellent tool for oilseed breeding, enabling the isolation of homozygous lines [106] and facilitating mutation-selection studies [107].

Microspore-derived embryos from both high and low erucic acid cultivars of *B. napus* (cultivars Reston and Topas, respectively) were shown to accumulate triacylglycerols in a manner and quantity similar to developing zygotic embryos in vivo [51]. The lipids found in late cotyledonary developing MD embryos of *B. napus* possessed an acyl composition very similar to that observed in lipids from mature zygotic embryos of the same cultivar (Table 8). In particular, MD Reston embryos were shown to accumulate oleic, eicosenoic and erucic acids during development, in a manner consistent with zygotic embryos in vivo (Table 9).

Analysis of mid-late cotyledonary stage MD embryos showed that *(Z)*-13 22:1 is excluded from the polar lipid fraction, but is present in the neutral lipids (DAGs and TAGs) as well as in the fatty acid and acyl-CoA pools [108]. Direct probe EI-MS analysis of endogenous lipid fractions indicated the presence of triacylglycerols containing (a) one eicosenoyl ($M^+ = 909–913$), (b) one erucoyl ($M^+ = 937–941$), (c) one erucoyl and one eicosenoyl ($M^+ = 965–969$) or (d) two erucoyl ($M^+ = 993–997$) moieties. The molecular ion clusters indicated combinations of very long chain fatty acids with 18:1, 18:2 or 18:3 in these triacylglycerols. Trierucoylglycerol (trierucin) ($M^+ = 1053$) was not found. The presence of erucoyl moieties at the *sn*-1 + 3 positions and the absence of this fatty acid at the *sn*-2 position of triacylglycerols was confirmed by analysis with pancreatic lipase. Thus, in general, the acyl composition of triacylglycerols isolated from developing MD embryos indicated that at the mid-late cotyledonary stage, eicosenoic and erucic acid were synthesized and accumulated in the neutral lipid fraction. Furthermore, the presence of 1,3-dierucoyl-2-acyl-*sn*-glycerol species would suggest that the enzymes responsible for incorporation of erucoyl moieties into both positions are present during MD embryo development [108].

Table 8. Content and fatty acid composition of triacylglycerols in mature microspore-derived embryos of *B. napus* cv. Reston and cv. Topas and in mature Reston and Topas seeds. Adapted from Taylor et al. [51]

B. napus cv.	Triacyl-glycerols (mg g^{-1} FW)	Fatty acid composition (mol%)[a]					
		16:0	18:1	18:2	18:3	20:1	22:1
Reston							
Cultured MD embryos	141	4	31	10	9	11	31
Seed embryos	408	4	21	13	7	11	40
Topas							
Cultured MD embryos	ND[b]	5	63	18	7		2[c]
Seed embryos	ND[b]	4	64	19	8		1[c]

[a] Minor proportions of various other fatty acids are not shown. [b] Not determined. [c] (20:1 + 22:1)

Table 9. Changes in fatty acid composition of total lipids with respect to proportions of oleic, eicosenoic, and erucic acid found in microspore-derived and zygotic embryos of *B. napus* cv Reston at different stages of development. Adapted from Taylor et al. [108]

Embryo system — Stage of Development	Fatty acid composition (Mole%)		
	18:1	20:1	22:1
Microspore-Derived			
Microspore	11.0	0.0	0.0
Heart	33.5	0.0	0.0
Torpedo	36.0	0.5	0.0
Early cotyledon	34.0	2.5	1.0
Mid cotyledon	44.0	7.0	5.0
Late cotyledon	38.0	8.0	11.0
Very late cotyledon	33.0	12.0	22.0
Zygotic			
Heart	18.0	0.0	0.0
Torpedo	21.0	1.0	1.0
Early cotyledon	25.0	2.0	0.5
Mid cotyledon	34.0	4.0	2.5
Late cotyledon	32.0	12.0	13.0
Very late cotyledon	27.0	12.0	26.5
Seed	24.0	11.5	33.0

4.3 In Vitro Studies of Triacylglycerol Bioassembly Using Cultured Embryos

4.3.1 Cultured Zygotic and Somatic Embryos

The few biochemical studies which have been conducted in cultured embryoids, will be reviewed here.

Turnham and Northcote [109] studied a tissue culture line of oil palm, *Elaeis guineensis*, producing embryoids which accumulated large quantities of lipids in the cells during embryogenesis. Using ^{14}C-acetate to monitor lipid accumulation, it was found that both triacylglycerol and polar lipid synthesis increased during embryoid formation. In the period before embryoids became visible, a rapid increase in the formation of polar lipids was noted, which corresponded to the increased rate of cell division and primary cell wall deposition occurring at the time [110]. The peak of incorporation of acetate into triacylglycerols or total lipids coincided with the time (day 35) at which embryoids became visible in the cultures [109] (Fig. 10) and was preceded by an increase in the activity of acetyl-CoA carboxylase (ACCase) in the tissue cultures. Thus, monitoring ACCase may be a useful way to detect embryogenesis in such cultures even before actual embryoids are visible.

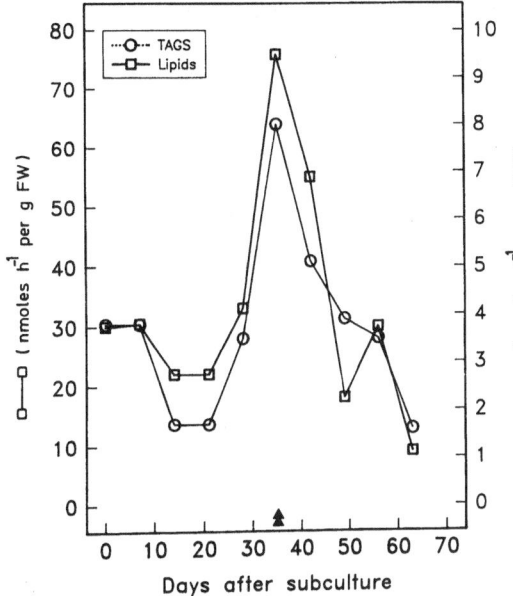

Fig. 10. Changes in the rate of incorporation of [1-^{14}C]acetate into total lipids (□) and triacylglycerols (○) during a time course of embryogenesis in oil palm tissue cultures. The data are presented as nmoles [1-^{14}C]acetate incorporated per hour into total lipids and triacylglycerols, respectively, on a fresh weight basis. Embryoids were visible in the cultures on day 35 *(arrow)*. After Turnham and Northcote [109]

In another study [111], protoplasts were enzymatically prepared from the mesocarp tissue of *E. guineensis* 16–20 weeks after anthesis and from rapidly multiplying embryogenic cultures from the same species. Radioactive incorporation studies showed that the protoplasts from both sources metabolized ^{14}C-acetate into all the main lipid classes, including triacylglycerols. However, the relative proportions of label in these classes varied widely, depending on the stage at which protoplasts were isolated from the mesocarp tissue or embryogenic culture, making it difficult to compare developmental accumulation of lipids in the two protoplast sources.

During somatic embryogenesis of *P. orientale*, incubation with ^{14}C-acetate resulted in its incorporation into acyl lipids. In particular, accumulation of radiolabel in triacylglycerols was maximal at about 4 weeks after induction of somatic embryogenesis [50]. This peak of acetate incorporation, representing de novo triacylglycerol assembly in vitro, coincides very closely with the peak of endogenous triacylglycerol accumulation reported in somatic embryo cultures of *Papaver* ssp. [49].

4.3.2 Microspore-Derived Embryos of *Brassica napus*

Since erucic acid (*(Z)*-13 22:1) is confined almost exclusively to the neutral lipid fraction in developing oilseeds [112], this fatty acid is an ideal marker for investigating enzymes involved in storage lipid biosynthesis. However, the mechanism for incorporation of *(Z)*-13 22:1 into triacylglycerols is not fully understood, despite several studies aimed at elucidating the pathway in various oilseeds [113–119]. This is, in part, due to the fact that, traditionally, metabolism studies conducted in zygotic oilseeds utilizing radiolabeled erucoyl moieties have shown very poor incorporation of this fatty acid into glycerolipids.

The *B. napus* microspore-derived embryo system has been investigated as a possible model for oilseed development, with a particular emphasis on seed-specific processes associated with the accumulation of storage lipids rich in erucic acid [51, 108, 118]. Preliminary studies [51] indicated that these embryos were able to metabolize erucoyl moieties very effectively in vitro. The presence of both erucoyl-CoA synthetase (E.C. 6.2.1.3) and erucoyl-CoA thioesterase (E.C. 3.1.2.2) activities was confirmed. In particular, the activity of the erucoyl-CoA synthetase was 5 to 20-fold higher than the corresponding activity monitored in preparations from zygotic oilseeds such as *Crambe abyssinica*, *Carthamus tinctorius* and *Cuphea wrightii* [115], or oleoyl-CoA formation in pea leaf microsomes [120].

Microspore-derived embryos from both early (14 d in culture) and mid-late (21–29 d in culture) cotyledonary stages yielded highly active cell-free enzyme preparations for studies of triacylglycerol biosynthesis in vitro [108].

Rates of triacylglycerol biosynthesis in vivo were estimated in both microspore-derived and zygotic embryos during the rapid phase of triacylglycerol accumulation. These rates were then compared with the capacity for homogenates from developing embryos to biosynthesize triacylglycerols in vitro using radiolabeled erucic acid as a marker (Table 10). While the maximal rates of triacylglycerol biosynthesis were essentially identical in the two embryo systems during embryogenesis in vivo, there was quite a contrast in in vitro capacities for triacylglycerol biosynthesis. The zygotic embryo homogenate exhibited a very low rate of incorporation of ^{14}C-erucoyl moieties into triacylglycerols in vitro. In contrast, the MD embryo homogenate was highly active in this regard, incorporating ^{14}C-erucic acid into triacylglycerols at rates which could more than support the maximal rate for triacylglycerol accumulation in vivo [108].

Table 10. Comparison of in vivo and in vitro rates of triacylglycerol biosynthesis in developing zygotic and microspore-derived embryos of *B. napus* cv Reston. After Taylor et al. [108]

Embryo system	Rate of TAG biosynthesis per dry weight $\mu g\ g^{-1}\ min^{-1}$	
	in vivo[a]	in vitro[b]
Zygotic	10.0	0.8 (8%)[c]
Microspore-derived	10.7	12.5 (117%)

[a] Estimated from measurements of TAG and dry weight in developing mid-late cotyledonary stage embryos during the rapid phase of TAG accumulation: Zygotic embryos, 4–8 weeks post anthesis; MD embryos, 1–5 weeks in culture. [b] Measured in homogenates from mid-late cotyledonary stage embryos using the reaction system: G-3-P + ^{14}C-erucoyl-CoA, and assuming one erucoyl moiety incorporated per TAG synthesized. [c] Values in parentheses express the in vitro rate as a percentage of the estimated in vivo rate

Table 11. Comparison of glycerolipids synthesized by an homogenate prepared from late cotyledonary stage microspore-derived embryos of *B. napus* cv Reston in the presence of 200 μM G-3-P and either 40 μM ^{14}C-oleoyl-CoA or ^{14}C-erucoyl-CoA. Adapted from Taylor et al. [108]

^{14}C-Acyl-CoA supplied	Incorporation of acyl moieties into various lipid species[a] (pmol min^{-1} per mg protein)						
	LPA/PA	MAG	DAG	TAG	FFA	LPC/PC	PE
Oleoyl-CoA	38	5	44	98	3	18	3
Erucoyl-CoA	nd[b]	nd	nd	170	32	nd	nd

[a] DAG, diacylglycerols; FFA, free fatty acids; LPA/PA, lysophosphatidic acids/phosphatidic acids; LPC/PC, lysophosphatidylcholines/phosphatidylcholines; MAG, monoacylglycerols; PE, phosphatidylethanolamines; TAG, triacylglycerols.
[b] nd = not detected

Homogenates incubated in the presence of G-3-P and [1-^{14}C]oleoyl-CoA produced radiolabeled PA, DAG and TAG as well as PC, PE and other complex polar lipids (Table 11). Radiolabeled 20:1 and 22:1 were not formed under these reaction conditions, although desaturation of 18:1 to 18:2 and 18:3 occurred to some extent. Such findings are consistent with the absence of exogenous malonyl-CoA in the reaction mixtures [117].

In marked contrast, in the presence of G-3-P and [1-^{14}C]erucoyl-CoA, homogenates were able to rapidly incorporate radiolabeled erucoyl moieties into triacylglycerols and the free fatty acid pool (Table 11). Labeled erucoyl moieties were not incorporated into other Kennedy pathway intermediates (LPA, PA, DAG) or complex polar lipids (e.g. PC, PE, MGDG, DGDG). The incorporation of ^{14}C-22:1 into triacylglycerols was linear over 30 min with up to 1 mg homogenate protein and was optimal in the presence of 200–500 μM G-3-P and 40 μM [1-^{14}C]erucoyl-CoA. This rate of ^{14}C-22:1 incorporation into triacylglycerols in the presence of G-3-P, exceeds those observed in homogenates or microsomal fractions of various developing zygotic oilseeds including *C. abyssinica* [115], *B. napus* [108, 114] and *Limnanthes douglasii* [119] by 30 to 80-fold (Fig. 11).

Chemicals known to inhibit phosphatidic acid phosphatase (25 mM EDTA) and diacylglycerol acyltransferase (5 mM DIPFP, 1.5 mM 2-BrOA) decreased the rate of incorporation of ^{14}C-erucoyl-CoA into triacylglycerols by 85%, 60% and 84%, respectively, but this inhibition was not accompanied by an accumulation of ^{14}C-erucoyl moieties in LPA, PA or DAG, or in other polar lipids [108].

G-3-P and a number of acyl lipids were tested as acceptors for ^{14}C-erucoyl-CoA. In general, all acceptors tested gave rates of triacylglycerol biosynthesis above that observed in the absence of exogenous acceptor. G-3-P, LPAs and DAGs were the best substrates. Analysis of radiolabeled triacylglycerols by HPLC indicated that, regardless of the acyl acceptor provided, only the triacylglycerol species containing labeled erucoyl moieties, i.e. 18:3/18:1/^{14}C-22:1, 18:2/18:1/^{14}C-22:1, 18:1/18:1/^{14}C-22:1 and 16:0/18:1/^{14}C-22:1, were formed in relative proportions of 12%, 30%, 55% and 3%, respectively [108, 118].

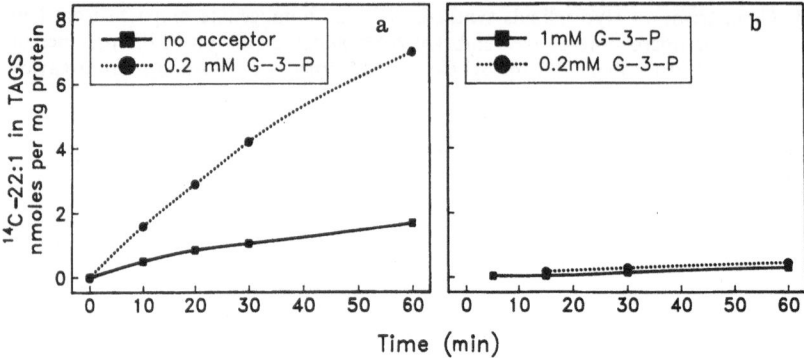

Fig. 11 a. Incorporation of ^{14}C-erucoyl moieties (supplied as 40 μM ^{14}C 22 : 1-CoA) into triacylglycerols by an homogenate of microspore-derived embryos of B. napus incubated in the presence (●) and absence (■) of G-3-P [108]. **b)** Incorporation of ^{14}C-erucic acid (19.2 μM) into triacylglycerols by an homogenate of developing seeds of *Crambe abyssinica*, incubated in the presence of 1 mM G-3-P (■) [115]; incorporation of ^{14}C G-3-P (200 μM) into triacylglycerols by a microsomal fraction from developing seeds of *Limnanthes douglasii* incubated in the presence of 100 μM 22 : 1-CoA (●) [119]

In the presence of 1,2-dierucin, in addition to the four triacylglycerol species containing erucoyl moieties, radiolabeled trierucin (22 : 1/22 : 1/^{14}C-22 : 1) was also produced and was confirmed by mass spectrometry [121] (Fig. 12). These results are of particular interest for genetic modification of rapeseed, e.g. by introducing foreign genes which may lead to very high erucic acid (>80%) cultivars.

A modified Brockerhoff stereospecific analysis of radiolabeled triacylglycerols produced in the presence of either ^{14}C-oleoyl-CoA or ^{14}C-erucoyl-CoA, indicated that radiolabeled erucoyl moieties were incorporated exclusively into the *sn*-3 position by a highly active diacylglycerol acyltransferase (E.C. 2.3.1.20), while radiolabeled oleoyl moieties were incorporated into the *sn*-1 and *sn*-2 positions on the glycerol backbone. Thus, while the radiolabeled incorporation data suggests that MD embryos possess all of the necessary enzymes of the Kennedy pathway [6, 108, 118] for triacylglycerol bioassembly, there appears to be some selectivity according to the acyl-CoA species supplied. The erucoyl-CoA: diacylglycerol acyltransferase activity has been localized primarily in a microsomal fraction obtained by differential centrifugation [108].

Studies of the biosynthesis of very long chain fatty acids and their incorporation into acyl lipids were conducted in early cotyledonary stage MD embryos of Reston, at the stage where very long chain fatty acids were beginning to accumulate (early cotyledonary stage). In the presence of malonyl-CoA, reducing equivalents and G-3-P, ^{14}C-oleoyl-CoA was elongated yielding radiolabeled eicosenoyl and erucoyl moieties in the triacylglycerol, fatty acid and acyl-CoA fractions. However, again ^{14}C-erucoyl moieties were excluded from other Kennedy pathway intermediates and complex polar lipids [118] (Table 12).

When tested in this elongation/incorporation system, the diacylglycerol acyltransferase inhibitor 2-BrOA caused rates of ^{14}C-oleoyl, ^{14}C-eicosenoyl, and

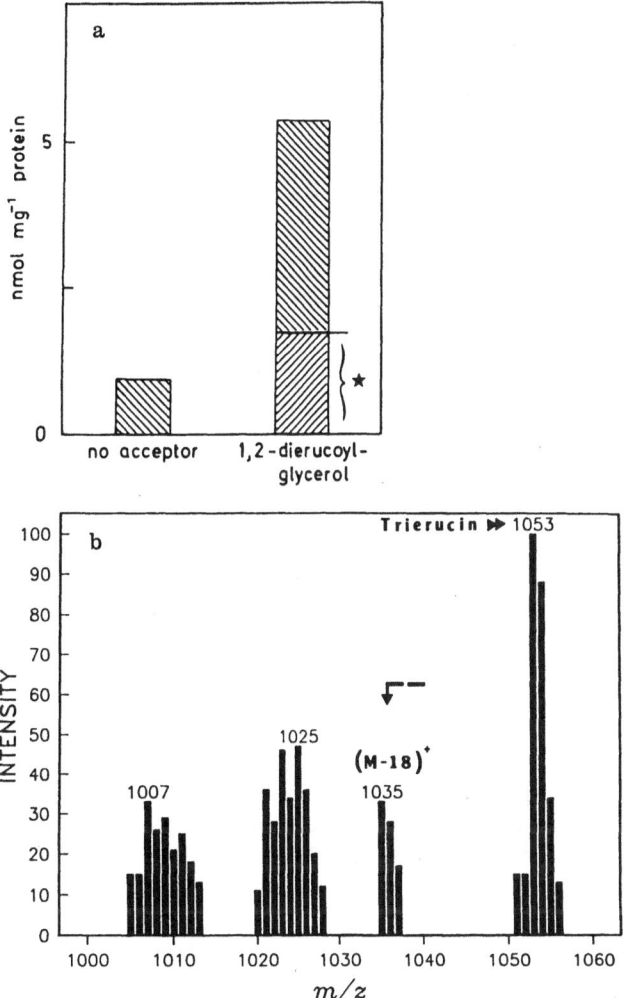

Fig. 12a. Incorporation of labeled erucoyl moieties into triacylglycerols by incubating an homogenate of cotyledonary stage microspore-derived embryos of *B. napus* L. cv Reston with [1-^{14}C]erucoyl-CoA as substrate and 1,2-dierucoylglycerol as acyl acceptor or without an exogenous acyl acceptor. * = Proportion of ^{14}C-trierucoylglycerol (~35%) in total labeled triacylglycerols. **b)** Partial molecular ion region of a direct probe electron impact-mass spectrum of triacylglycerols biosynthesized as described in (*a*). Adapted from Taylor et al. [121]

[14]C-erucoyl incorporation into triacylglycerols to decrease by 60%, 80% and 100%, respectively. There was a concomitant increase in incorporation of ^{14}C-oleoyl moieties into LPA/PA and DAG but these increases were not accompanied by the appearance of radiolabeled eicosenoyl or erucoyl moieties in these Kennedy pathway intermediates (Table 12). Thus, at present, the mechanism

Table 12. Biosynthesis of lipid species containing ^{14}C-labeled oleoyl, eicosenoyl, and erucoyl moieties by homogenates prepared from early cotyledonary stage (14 d in culture) microspore-derived embryos. Incubations were conducted under various elongation/incorporation conditions. Final concentrations of G-3-P, acyl-CoAs and malonyl-CoA were 200 µM, 18 µM and 1 mM, respectively. After Taylor et al. [118]

Reaction components	^{14}C-Acyl moieties	Incorporation of ^{14}C-acyl moieties into lipid species[a] (pmol h^{-1} per mg protein)				
		LPA/PA	DAG	TAG	FFA	PC
[14]C-Oleoyl-CoA + G-3-P	18:1[b]	546	1164	2256	174	870
[14]C-Oleoyl-CoA + malonyl-CoA + NADH, NADPH + G-3-P	18:1 20:1 22:1	204 0 0	576 42 0	2946 497 222	294 66 48	222 0 0
[14]C-Oleoyl-CoA + malonyl-CoA + NADH, NADPH + G-3-P + 2-BrOA[c]	18:1 20:1 22:1	552 0 0	828 0 0	1139 108 0	438 60 0	690 0 0
[14]C-Erucoyl-CoA + G-3-P	22:1[d]	0	0	4830	275	0
[14]C-Erucoyl-CoA + G-3-P + 2-BrOA[c]	22:1	0	0	1720	1746	0

[a] DAG, diacylglycerols; FFA, free fatty acids; LPA/PA, lysophosphatidic acids/phosphatidic acids; PC, phosphatidylcholines; TAG, triacylglycerols. [b] In the absence of malonyl-CoA and/or reducing equivalents no elongation of ^{14}C-oleoyl was observed. [c] 2-Bromo-octanoate (2-BrOA) supplied at a final concentration of 1.5 mM. [d] No other ^{14}C-acyl moieties were detected

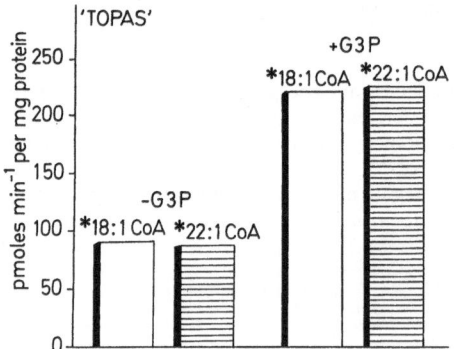

Fig. 13. Incorporation of ^{14}C-oleoyl-CoA and ^{14}C-erucoyl-CoA, 40 µM each, into triacylglycerols by homogenates of microspore-derived embryos of *B. napus* cv Topas (low erucic acid cultivar) incubated in the presence and absence of 200 µM G-3-P. The data are presented as pmoles labeled acyl moieties incorporated per min and mg protein. D. C. Taylor (unpublished data)

by which erucoyl moieties are incorporated into the sn-1 position of triacylglycerols is unclear [118].

Microspore-derived embryos from a low erucic acid *B. napus* cultivar (Topas) were also screened for the ability to biosynthesize triacylglycerols in vitro. In the presence of G-3-P, homogenates prepared from mid-cotyledonary stage MD Topas embryos incorporated erucoyl- and oleoyl moieties into triacylglycerols at equal rates (Fig. 13). Thus, the breeding effort which produced low erucic acid cultivars of *B. napus* did not adversely affect the capacity of such embryos to incorporate very long-chain fatty acids into triacylglycerols. Rather, such cultivars are impaired in the ability to biosynthesize eicosenoic and erucic acids.

5 Conclusions

Advances in plant biotechnology and genetic engineering have stimulated the application of cell culture methods to the breeding of agriculturally important oilseed crops and other plants. For example, studies aimed at developing improved oil crops may be carried out more efficiently by utilizing the biosynthetic capacity of cultured plant cells. The pathway leading to the formation of triacylglycerols in plant cell cultures is apparently not different from that prevalent in developing seeds. Biosynthesis of appreciable proportions of triacylglycerols and various unusual fatty acids, however, seems to be correlated with embryogenesis in many plants. Obviously, the genetic control of the biosynthesis of triacylglycerols and unusual fatty acids in plant cell cultures comprises, at least, two different sets of developmental signals which are of particular interest for utilization of such embryogenic cells as models in plant breeding.

Recent studies on developing somatic and gametophytic embryos of oil crops show that embryo formation in culture parallels embryogenesis in developing seeds in many respects, e.g. storage lipid assembly and fatty acid elongation. Above all, the microspore-derived embryo system, e.g. of *B. napus*, is an attractive alternative to zygotic embryos for studies of lipid biosynthetic enzymes. It is easy to obtain specific developmental stages which exhibit high capacities for storage lipid biosynthesis and accumulation in vitro. Furthermore, it is evident that cultured gametophytic embryos of plants, in particular microspore-derived (androgenic) embryos, forming non-endospermic oilseeds may be useful in studies concerned with the improvement of oil plants. They may facilitate and speed up the selection and multiplication of clones of superior quality produced, for example, by unconventional breeding methods.

6 Nomenclature and Abbreviations

Lipids are named according to the 1976 recommendations of the IUPAC-IUB Commission on Biochemical Nomenclature [Eur. J. Biochem. 79: 1 (1977)]. Fatty acids are characterized by number of carbon atoms: number of double bonds, e.g. 16:0, palmitic acid; *(Z)*-9 18:1, oleic acid [*(Z)*-9-octadecenoic acid]; *(Z)*-13

22:1, erucic acid [*(Z)*-13-docosenoic acid], etc. The geometry of double bonds of fatty acids is indicated by the prefixes *(Z)* and *(E)* instead of *cis* and *trans*, respectively, following IUPAC-IUB recommendations. The carbon atoms of the glycerol backbone are designated by stereospecific numbering, *sn*, according to IUPAC-IUB recommendations.

Acyl-CoA acyl-Coenzym A, e.g. *(Z)*-9 18:1-CoA, oleoyl-Coenzyme A (see above)
AT   acyltransferase
2-BrOA   2-bromooctanoic acid
2,4-D   2,4-dichlorophenoxy acetic acid
DAG   diacylglycerols
DGDG   digalactosyl diacylglycerols
DIPFP   diisopropyl fluorophosphate
DW   dry weight
EDTA   ethylenediamine tetraacetic acid
EI-MS   electron impact-mass spectrometry
FA   fatty acid
FW   fresh weight
G-3-P   glycerol-3-phosphate
HPLC   high performance liquid chromatography
LPA   2-lysophosphatidic acids
MD   microspore-derived
MGDG   monogalactosyl diacylglycerols
NADH   nicotinamide adenine dinucleotide (reduced)
NADPH   nicotinamide adenine dinucleotide phosphate (reduced)
PA   phosphatidic acids
PC   phosphatidylcholines
PE   phosphatidylethanolamines
P_i inorganic phosphate
PL   polar lipids
TAG   triacylglycerols
TL   total lipids
Tr   trace
VLCFA   very long-chain fatty acids (C $\geq$ 20)

7 References

1. Knauf VC (1987) Trends Biotechnol 5: 40
2. Eckes P, Donn G, Wengenmayer F (1987) Angew Chem 99: 392
3. Röbbelen G (1987) In: Applewhite TM (ed) Proceedings world conference on biotechnology for the fats and oils industry. American Oil Chemists' Society, Champaign, IL, p 78
4. Somerville CR, Browse J (1988) In: Conn EE (ed) Recent advances in phytochemistry, vol 22. Plenum, New York, p 19
5. Battey JF, Schmid KM, Ohlrogge JB (1989) Trends Biotechnol 7: 122
6. Stymne S, Stobart AK (1987) In: Stumpf PK, Conn EE (eds-in-chief) The biochemistry of plants, vol 9. Academic, New York, p 175

7. Gurr MI (1984) In: Stumpf PK, Conn EE (eds-in-chief) The biochemistry of plants, vol 4. Academic, New York, p 205
8. Ohlrogge JB (1987) In: Applewhite TH (ed) Proceedings world conference on biotechnology for the fats and oil industry. American Oil Chemists' Society, Champaign, IL, p 87
9. Roughan PG, Slack CR (1982) Annu Rev Plant Physiol 33, 97
10. Yermanos DM (1975) J Am Oil Chem Soc 52: 115
11. Kolattukudy PE (1984) In: Stumpf PK, Conn EE (eds-in-chief) The biochemistry of plants, vol 4. Academic, New York, p 571
12. Radwan SS, Mangold HK (1976) Adv Lipid Res 14: 171
13. Radwan SS, Mangold HK (1980) In: Fiechter A (ed) Advances in biochemical engineering, vol 16. Springer, Berlin Heidelberg New York, p 109
14. Kleinig H, Kopp C (1978) Planta 139: 61
15. Kleinig H, Steinki C, Kopp C, Zaar K (1978) Planta 140: 233
16. Radwan SS, Grosse-Oetringhaus S, Mangold HK (1978) Chem Phys Lipids 22: 177
17. Kleinig H, Hara S, Schuchmann R (1982) In: Fujiwara (ed) Plant tissue culture 1982, Maruzon, Tokyo, p 257
18. Theimer RR, Schöpf UFM (1989) Fat Sci Technol 91: 434
19. Schöpf UFM, Theimer RR (1980) Biol Chem Hoppe-Seyler 370: 798
20. Tsai CH, Wen MC, Kinsella JE (1982) J Food Sci 47: 768
21. Tsai CH, Kinsella JE (1981) Lipids 16: 577
22. Foster C, End M, Leathers R, Pettipher G, Hadley P, Scragg AH (1988) In: Applewhite TH (ed) Proceedings world conference on biotechnology for the fats and oil industry. American Oil Chemists' Society, Champaign, IL, p 305
23. Song M, Tattrie N (1973) Can J Bot 51: 1893
24. Radwan SS (1976) Phytochemistry 15: 1727
25. Schneiders G (1981) Zusammensetzung und Stoffwechsel von Lipiden in photosynthetisch aktiven Zellsuspensionskulturen, Thesis, University of Münster
26. Radwan SS (1975) Fette Seifen Anstrichm 77: 181
27. Wilson AC, Kates M, de la Roche AI 81978) Lipids 13: 504
28. Martin BA, Horn ME, Widholm JM, Rinne RW (1984) Biochim Biophys Acta 796: 146
29. Ellenbracht F, Barz W, Mangold HK (1980) Planta 150: 114
30. Leathers RR, Scragg AH (1989) Plant Sci 62: 217
31. Scragg AH, Leathers RR (1988) In: Moreton RS (ed) Single cell oil. Longman, Harlow, England, p 71
32. Gregor HD (1977) Chem Phys Lipids 20: 77
33. Manoharan K, Prasad R, Guha-Mukherjee S (1990) Phytochemistry 29: 2529
34. Ohyama K, Uchida Y, Misawa N, Komano T, Fujita M, Ueno T (1984) Plant Cell Rep 3: 21
35. Mangold HK (1977) In: Barz W, Reinhard E, Zenk MH (eds) Plant tissue culture and its bio-technological application. Springer, Berlin Heidelberg, p 55
36. Halder T, Gadgil VN (1983) Phytochemistry 22: 1965
37. Ezzat KS, Pearce RS (1980) Phytochemistry 19: 1375
38. Gemmrich AR (1982) Plant Cell Rep 1: 233
39. Dutta PC, Appelqvist L-Å (1989) Plant Sci 64: 167
40. Radwan SS, Spener F, Mangold HK, Staba EJ (1975) Chem Phys Lipids 14: 72
41. Belay S, Rier JP, Ayorinde FO (1989) J Am Oil Chem Soc 66: 828
42. Yano I, Nichols BW, Morris LJ, James AT (1972) Lipids 7: 30
43. Radwan SS, Mangold HK, Spener F (1974) Chem Phys Lipids 13: 103
44. Radwan SS, Mangold HK, Hüsemann W, Barz W (1979) Chem Phys Lipids 24: 79
45. Staba EJ, Shik Shin B, Mangold HK (1971) Chem Phys Lipids 6: 291
46. Tattrie NH, Veliky IA (1973) Can J Bot 51: 513
47. Spener F, Staba EJ, Mangold HK (1974) Chem Phys Lipids 12: 344
48. Gemmrich AR, Schraudolf H (1980) Chem Phys Lipids 26: 259
49. Schuchmann R, Wellmann E (1983) Plant Cell Rep 2: 88
50. Hara S, Falk H, Kleinig H (1985) Planta 164: 303

51. Taylor DC, Weber N, Underhill EW, Pomeroy MK, Keller WA, Moloney MM, Wilen RW, Scowcroft WR, Holbrook LA (1990) Planta 181: 18
52. Falkenau C, Heim S, Wagner KG (1987) Plant Sci 50: 173
53. Connett RJA, Hanke DE (1987) Planta 170: 161
54. Ettlinger C, Lehle L (1988) Nature 331: 176
55. Finkelstein RR, Tenbarge KM, Shumway JE, Crouch M (1985) Plant Physiol 78: 630
56. Kates M, Wilson AC, de la Roche AJ (1979) In: Appelqvist L-Å, Liljenberg C (eds) Advances in the biochemistry and physiology of plant lipids. Elsevier/North-Holland, Amsterdam, p 329
57. Tsai CH, Kinsella JE (1982) Lipids 17: 367
58. Tsai CH, Kinsella JE (1982) Lipids 17: 848
59. Stumpf PK, Weber N (1977) Lipids 12: 120
60. Weber N, Richter I, Mangold HK, Mukherjee KD (1979) Planta 145: 479
61. De Silva NS, Fowler MW (1976) Phytochemistry 15: 1735
62. Weber N, Mangold HK (1983) Planta 158: 111
63. Weber N, Benning H (1985) Eur J Biochem 146: 323
64. Weber N, Mangold HK (1988) In: Constabel F, Vasil IK (eds) Cell culture and somatic cell genetics of plants, vol 5, Academic, New York, p 509
65. Thies W (1971) Fette Seifen Anstrichm 73: 710
66. Tisserat B, Esan EB, Murashige T (1979) Hort Rev 1: 1
67. Janick J (1986) In: Crocomo OJ, Sharp WR, Evans DA, Bravo JE, Tavares FCA, Paddock EF (eds) Biotechnology of plants and microorganisms. Chapter 8, Ohio State University, Columbus, p 97
68. Fan Z, Armstrong KC, Keller WA (1988) Protoplasma 147: 191
69. Maheshwari P, Sachar RC (1963) In: Maheshwari P (ed) Recent advances in the embryology of angiosperms. International Society of Plant Morphologists, University of Dehli, Dehli, India, p 265
70. Yang HY, Zou C (1982) Theor Appl Genet 63: 87
71. Avjioglu A, Knox RB (1989) Ann Bot 63: 409
72. Norton G, Harris JF (1975) Planta 123: 163
73. Norton G, Harris JF (1983) Phytochemistry 22: 2703
74. Zhang D, Zhengua C, Lihua Z, Wenbin L (1988) Acta Gent Sinica 15: 254
75. Janick J, Simon JE, Quinn J, Beaubaire N (1989) In: Craker LE, Simon JE (eds) Herbs, spices and medicinal plants: Recent advances in botany, Horticulture, vol 4, Oryx, Phoenix, p 145
76. Whipkey A, Simon JE, Janick J (1988) J Am Oil Chem Soc 65: 979
77. Quinn J, Whipkey A, Simon J, Janick J (1987) Acta Hort 208: 243
78. Pence VC, Hasegawa PM, Janick J (1981) J Am Soc Hort Sci 106: 381
79. Pence VC, Hasegawa PM, Janick J (1981) Physiol Plant 53: 378
80. Pence VC, Hasegawa PM, Janick J (1980) Z. Pflanzenphysiol. 98: 1
81. Janick J, Wright DC, Hasegawa PM (1982) J Am Soc Hort Sci 107: 919
82. Kononowicz AK, Janick J (1984) J Am Soc Hort Sci 109: 266
83. Wright DC, Kononowicz AK, Janick J (1984) J Am Soc Hort Sci 109: 77
84. Wright DC, Janick J, Hasegawa PM (1984) Lipids 18: 863
85. Miwa TK (1971) J Am Oil Chem Soc 48: 259
86. Wang Y-C, Janick J (1986) J Am Soc Hort Sci 111: 797
87. Cherry JH, Bishop L, Leopold N, Pikaard C, Hasegawa PM (1984) Phytochemistry 23: 2183
88. Shoemaker RC, Hammond EG (1988) In Vitro Cell Develop Biol 24: 829
89. Rabechault H, Martin JP, Cas S (1972) Oleagineux 27: 318
90. Jones LH (1974) In: Spencer B (ed) Industrial aspects of biochemistry, Elsevier/North-Holland, Amsterdam, p 813
91. Rabechault H, Martin JP (1976) CR Hebd Seances Acad Sci Paris 283: 1735
92. Jones LH, Hughes WA (1988) In: Bajaj YPS (ed) Biotechnology in agriculture and forestry, vol 5: Trees II. Springer, Berlin, p 176
93. Jones LH (1989) Biotechnol Genet Eng Rev 7: 281

94. Schwendiman J, Pannetier C, Michaux-Ferriere N (1988) Ann Bot 59: 43
95. Jones LH (1984) J Am Oil Chem Soc 61: 1717
96. Kudielka RA, Theimer RR (1983) Plant Sci Lett 31: 237
97. Kudielka RA, Theimer RR (1983) In: Proceedings workshop space biology. ESA SP-206: 63 (from: Theimer RR, Kudielka RA, Rösch I (1986) Naturwissenschaften 73: 442)
98. Kleiman R, Spencer GF (1982) J Am Oil Chem Soc 59: 29
99. Matsuzaki T, Iwai S, Koiwai A (1988) Agric Biol Chem 52: 1283
100. Thomas E, Wenzel G (1975) Z Pflanzenzücht 74: 77
101. Keller WA, Armstrong KC (1977) Can J Bot 55: 1383
102. Wenzel G, Hoffmann F, Thomas E (1977) Z Pflanzenzücht 78: 149
103. de la Roche AI, Keller WA (1977) Z Pflanzenzücht 78: 319
104. Lichter R (1982) Z Pflanzenzücht 105: 427
105. Chuong PV, Beversdorf WD (1985) Plant Sci 39: 219
106. Keller WA, Armstrong KC, de la Roche IA (1982) In: Giles KL, Sen SK (eds) Plant cell culture in crop improvement. Plenum, New York, p 169
107. Polsoni L, Kott LS, Beversdorf WD (1988) Can J Bot 66: 1681
108. Taylor DC, Weber N, Barton DL, Underhill EW, Hogge LR, Weselake RJ, Pomeroy MK (1991) Plant Physiol 97: 65
109. Turnham E, Northcote DH (1984) Phytochemistry 23: 35
110. Turnham E, Northcote DH (1982) Biochem J 208: 323
111. Sambanthamurthi R, Oo K-C, Ong ASH (1987) Plant Sci 51: 97
112. Roughan PG, Slack CR (1982) Annu Rev Plant Physiol 33: 97
113. Mukherjee KD (1986) Planta 167: 279
114. Sun C, Cao Y-Z, Huang AHC (1988) Plant Physiol 88: 56
115. Battey JF, Ohlrogge JB (1989) Plant Physiol 90: 835
116. Fehling E, Mukherjee KD (1990) Phytochemistry 29: 1525
117. Agrawal VP, Stumpf PK (1985) Lipids 20: 361
118. Taylor DC, Weber N, Barton D, Underhill E, Pomeroy K (1990) In: Quinn PJ, Harwood JL (eds) Plant lipid biochemistry, structure and utilization. The Biochemical Society, Portland, London, p 210
119. Cao Y-Z, Oo K-C, Huang AHC (1990) Plant Physiol 94: 1199
120. Murphy DJ, Mukherjee KD, Latzko E (1983) Biochem J 213: 249
121. Taylor DC, Weber N, Hogge LR, Underhill EW, Pomeroy MK, (1992) J Am Oil Chem Soc (in press)

Author Index Volumes 1−45

Errata to Vol. 44

Contribution: V. Kren, "Bioconversions of Ergot Alkaloids"

p. 125, the correct Fig. 2 is:

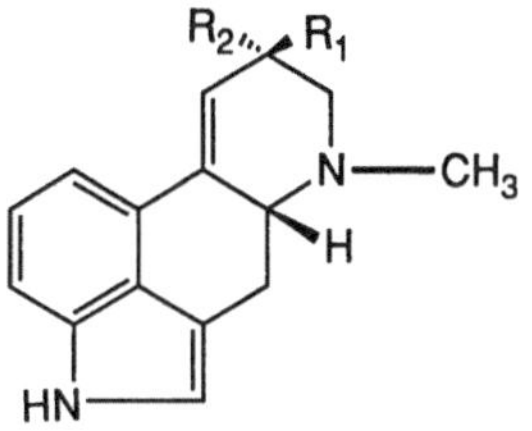

Fig. 2. 9,10-Ergolenes, lysergic acid derivatives
(5) lysergic acid R_1 = COOH, R_2 = H
(6) isolysergic acid R_1 = H, R_2 = COOH
(7) ergine R_1 = CONH$_2$, R_2 = H
(8) erginine R_1 = H, R_2 = CONH$_2$
(9) ergometrine R_1 = CONHCH(CH$_3$)CH$_2$OH,
$\qquad R_2$ = H
(10) lysergic acid α-hydroxyethylamide
$\qquad R_1$ = CONHCH(CH$_3$)OH, R_2 = H
(11) lysergol R_1 = CH$_2$OH, R_2 = H
(12) iso-lysergol R_1 = H, R_2 = CH$_2$OH
(13) lysergene $R_{1,2}$ = CH$_2$
(14) lysergine R_1 = CH$_3$, R_2 = H
(15) setoclavine R_1 = CH$_3$, R_2 = OH
(16) penniclavine R_1 = CH$_2$OH, R_2 = OH
(17) isosetoclavine R_1 = OH, R_2 = CH$_3$
(18) isopenniclavine R_1 = OH, R_2 = CH$_2$OH
(29) 8-hydroxyergine R_1 = CONH$_2$, R_2 = OH

p. 137, the correct Fig. 9 is:

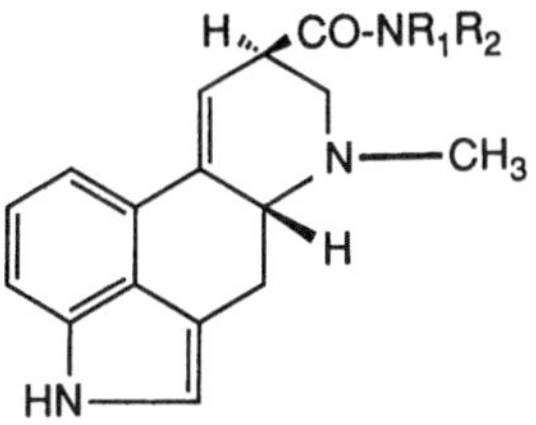

Fig. 9. LSD and its metabolites
(33) lysergic acid diethylamide (LSD)
$\qquad R_1$ = R_2 = CH$_2$CH$_3$
(34) lysergic acid ethylamide R_1 = H,
$\qquad R_2$ = CH$_2$CH$_3$
(35) lysergic acid vinylamide R_1 = H,
$\qquad R_2$ = CH =CH$_2$
(36) lysergic acid ethyl 2-hydroxyethylamide
$\qquad R_1$ = CH$_2$CH$_3$, R_2 = CH$_2$CH$_2$OH